198
Advances in Polymer Science

Advances in Polymer Science
Recently Published and Forthcoming Volumes

Surface-Initiated Polymerization II

Volume Editor: Rainer Jordan

With contributions by

B. Akgun · M. Baum · D. E. Bergbreiter · R. R. Bhat · C. Blickle
S. G. Boyes · W. J. Brittain · M. D. Foster · J. Genzer · A. M. Granville
A. M. Kippenberger · B. K. Mirous · A. Naji · R. R. Netz · C. Seidel
M. R. Tomlinson · T. Wu · B. Zhao

 Springer

The series *Advances in Polymer Science* presents critical reviews of the present and future trends in polymer and biopolymer science including chemistry, physical chemistry, physics and material science. It is adressed to all scientists at universities and in industry who wish to keep abreast of advances in the topics covered.
As a rule, contributions are specially commissioned. The editors and publishers will, however, always be pleased to receive suggestions and supplementary information. Papers are accepted for *Advances in Polymer Science* in English.
In references *Advances in Polymer Science* is abbreviated *Adv Polym Sci* and is cited as a journal.

Springer WWW home page: http://www.springer.com
Visit the APS content at http://www.springerlink.com/

Library of Congress Control Number: 2005935449

ISSN 0065-3195
ISBN-10 3-540-30251-4 Springer Berlin Heidelberg New York
ISBN-13 978-3-540-30251-3 Springer Berlin Heidelberg New York
DOI 10.1007/11586142

Springer is a part of Springer Science+Business Media

springer.com

The use of registered names, trademarks, etc. in this publication does not imply, even in the absence of a specific statement, that such names are exempt from the relevant protective laws and regulations and therefore free for general use.

Cover design: *Design & Production* GmbH, Heidelberg
Typesetting and Production: LE-TeX Jelonek, Schmidt & Vöckler GbR, Leipzig

Printed on acid-free paper   02/3100 YL – 5 4 3 2 1 0

Advances in Polymer Science
Also Available Electronically

For all customers who have a standing order to Advances in Polymer Science, we offer the electronic version via SpringerLink free of charge. Please contact your librarian who can receive a password or free access to the full articles by registering at:

springerlink.com

If you do not have a subscription, you can still view the tables of contents of the volumes and the abstract of each article by going to the SpringerLink Homepage, clicking on "Browse by Online Libraries", then "Chemical Sciences", and finally choose Advances in Polymer Science.

You will find information about the

– Editorial Board
– Aims and Scope
– Instructions for Authors
– Sample Contribution

at springer.com using the search function.

Preface

These two volumes on surface-initiated polymerization deal with recent developments in the synthesis, characterization and properties of structurally and chemically defined polymer coatings on surfaces. Nearly all polymerization techniques that have been developed in solution have now been adapted for the surface-initiated polymerization (SIP). The reader will find all relevant techniques discussed in these volumes, such as free, controlled and living radical polymerization, living anionic and cationic polymerization (Rigoberto Advincula), and ring-opening metathesis polymerization (Michael Buchmeiser). Most of them are used to prepare so-called polymer brushes, a term describing strictly linear polymers that are densely grafted via one end to an interface. Such coatings display unique physical properties useful for a variety of applications. In particular, the high structural control of polymer brushes that can be realized by controlled or living polymerization techniques draws much attention. The contribution by Takeshi Fukuda et al. on high-density polymer brushes outlines the synthetic possibilities as well as the unique properties of polymer brushes. Such coatings will surely play an important role in innovative surface science and nanotechnology. The present contributions also reflect an ongoing trend: the development of defined heterogeneities on nearly any length scale. This can be realized by structured polymer coatings, gradients and control of the topography via the SIP reaction conditions. Jan Genzer's contribution on the preparation of polymer brush gradients is a good example. As it relates to defined structural variation and control of the macromolecular design of grafting polymers via SIP, I would like to point the reader to the contributions by Takehisa Matsuda on surface graft microachitectures or by David Bergbreiter discussing the synthesis and applications of hyperbranched polymers on surfaces.

Originally, the reviews were to be divided into, e.g., a *Synthesis*, *Properties* and *Application* section. Fortunately, this was not possible at all. Synthesizing a polymer coating by SIP is performing materials science from scratch. Introducing a slightly different monomer or changing the solvent will automatically alter the properties of the surface such as its wetting behavior, topography, elasticity, homogeneity, etc. It is exciting (and difficult!) to characterize the layers and find out why an altered reaction condition had such an impact upon the various layer properties. Thus, the researcher is immediately involved in various aspects of surface science and analytical challenges. This is reflected in all contributions. For example, Daniel Dyer discusses the fundamental and interesting aspect of the photoinitiated synthesis of polymer brushes. Of course, the enormous advances in surface-sensitive characterization techniques developed for the investigation of self-assembled monolayers have provided the proper tools. However, as polymers are flexible, the investigation of the dynamic behavior of polymer coatings adds another dimension. The contribution by William Brittain on stimuli-responsive films gives an idea of the complex behavior of polymer brushes.

Besides the analytical techniques, the theoretical description of polymer brushes allows a deeper understanding of the complex dynamic behavior of polymers on surfaces and is useful for future developments. Here, Roland Netz gives – also for the non-expert – a very helpful theoretical background on the theoretical approaches for the description of neutral and charged polymer brushes.

The interest in polymer brushes and defined polymer coatings prepared via SIP is not at all restricted to the polymer community or the surface science community. The demand for tailored, functionalized and adaptive surfaces comes from a multitude of scientific branches and also from industry. Possible applications are already discussed in many of the contributions compiled here. Besides polymer science, surface chemistry and physics, they include catalysis, biomedical applications, microfluidics and nanotechnology. This creates a highly interdisciplinary, lively and fruitful environment.

Finally, I would like to thank all authors for their time and effort to make a state-of-the-art overview of surface-initiated polymerization possible. An edited book is only as good as its contributions and I had the privilege to compile contributions of the highest quality.

I am also grateful to Ms. Ulrike Kreusel and Dr. Marion Hertel from Springer for their professional help and patience.

Munich, January 2006   *Rainer Jordan*

Contents

Contents of Volume 197

Surface-Initiated Polymerization I

Volume Editor: Rainer Jordan
ISBN: 3-540-30247-6

Adv Polym Sci (2006) 198: 1–49
DOI 10.1007/12_059
© Springer-Verlag Berlin Heidelberg 2006
Published online: 19 January 2006

Hyperbranched Surface Graft Polymerizations

David E. Bergbreiter (✉) · Andrew M. Kippenberger

Department of Chemistry, College Station, Texas A&M University,
Texas, TX 77842-3012, USA
bergbreiter@tamu.edu

Abstract This review summarizes the synthesis of irregular hyperbranched polymer grafts on various inorganic and organic substrates. The synthesis of these hyperbranched grafts are generally based on "graft on a graft" polymerizations and include diverse sorts of graft polymers. The "graft-on-a-graft" strategies discussed here include chemistry leading to the synthesis of hyperbranched poly(acrylic acid) grafts, polysiloxane grafts, dendrimer/polyanhydride graft nanocomposites, ring-opening polymerization grafts, and polyamidoamine grafts. Other relevant chemistry of these grafts including chemistry leading to derivatives of hyperbranched poly(acrylic acid) grafts, further modification by polyionic interactions, polyvalent hydrogen bonding, and functional group manipulation is discussed. Examples of reactions of monomers with polyvalent surfaces that lead to hyperbranched grafts are also briefly discussed.

Keywords Dendrimer · Hyperbranched grafts · Nanocomposite · Polyvalency · Surface modification

Abbreviations

PAA   poly(acrylic acid)
PTBA   poly(*tert*-buyl acrylate)
MUA   mercaptoundecanoic acid
FTIR-ERS Fourier Transform Infrared external reflection spectroscopy
XPS   X-ray photoelectron spectroscopy
PE   polyethylene
PP   polypropylene
ATR-IR   attenuated total reflectance infrared
PNIPAM   poly(*N*-isopropylacrylamide)
TEA   2-thiopheneethyleneamine
ROP   ring opening polymerization
APES   3-aminopropyltriethoxysilane

1
Introduction

There is great interest in designing functional interfaces. Hyperbranched grafts are alternatives to existing "linear" grafts for formation of such interfaces. They are of interest because they can provide interfaces with different sorts of properties. Hyperbranched grafting is also conceptually more attractive than other approaches because the multiple grafting of oligomeric grafting reagents can compensate for inefficiencies in reactions at surfaces (Fig. 1). If, for example, an initial surface graft has coverage defects or if defects are introduced during the graft-on-a-graft synthesis due to incomplete reactions, subsequent hyperbranched grafting stages can "heal" these defects more efficiently than the traditional monomer grafting strategies that produce linear graft chains (Fig. 1b versus 1a). This same effect was noted previously by Ferguson in layer-by-layer grafting of mica particles and polycationic polymers on hydrophobic surfaces like octadecyltrichlorosilane treated Si/SiO₂ wafers and hexadecanethiol-modified silver films and is a general feature common to other layer-by-layer grafting chemistry [1, 2]. As shown in Fig. 1, the advantages of hyperbranching are considerable. In the particular schematic drawing of three graft stages shown in Fig. 1b, hyperbranched grafting is far more effective than linear grafting through three stages in Fig. 1a even when there is a relatively low (50%) efficiency in the first step of grafting.

The synthetic strategies that lead to irregularly hyperbranched grafts based on surface confined "graft-on-a-graft" polymerization reactions are the focus of this review. Limited examples of monomers reacting with polyvalent surface-bound reagents leading to hyperbranched polymers are also discussed. In general, the chemistry described here is confined to reactions

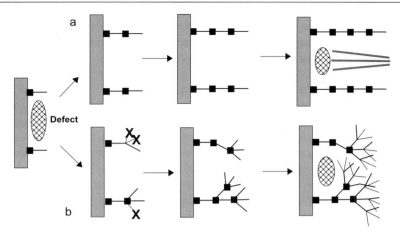

Fig. 1 A schematic drawing comparing linear grafting **a** with hyperbranched grafting **b** in coverage or "healing" of surface defects. An efficiency of 100% is assumed in all three steps in linear grafting (**a**). In the hyperbranched graft example (**b**), a 50% efficiency is assumed in the first step but 100% efficiency and three branches per graft stage are assumes in steps two and three

that involve condensation polymerization reactions or reactions that involve the reaction of an electrophile with a nucleophile. This review begins with hyperbranched grafting of poly(acrylic acid) on hard inorganic or metal surfaces and soft polymer surfaces. Methods for derivatizing these films either by covalent modification or with polyvalent noncovalent interactions are discussed. Limited examples of applications of these materials are described. For example, Crooks' group has used some of these synthetic methods to prepare patterned surfaces. In cases like this where this subject has been reviewed, it is only briefly discussed here. Other hyperbranched grafting strategies including multilayer grafting of polyvalent nucleophiles and electrophiles, grafting via ring opening polymerizations, and the synthesis of dendritic grafts using polyvalent surface-bound reagents and monomers are discussed subsequently. There are other very successful synthetic strategies for preparing hyperbranched films based on free radical polymerizations that will not be a topic of discussion in this review. For example, Müller has developed a novel method of hyperbranched graft polymerization of inimers (initiator-monomers) by self-condensed vinyl polymerization (SCVP) via atom transfer polymerization (ATRP) [3, 4]. Another example would be Matsuda's preparation of hyperbranched grafts by iniferter (initiator-transfer agent-terminator) polymerization [5, 6]. A detailed description of these iniferter polymerizations can be found in Matsuda's contribution in this volume. A similar approach by Tsubokawa is described as a post-graft polymerization of vinyl monomers and is useful as a route to hyperbranched grafts [7–9]. Surfaces with hyperbranched grafts can also be prepared by grafting commercially available hyperbranched polymers to surfaces. For example, Tsukruk has studied grafted hyperbranched polyesters with terminal epoxides that are attached to Si – OH surfaces [10, 11]. There are many examples where dendrimers are attached to surfaces by covalent or non-covalent interactions [12–17]. This chemistry too is not discussed here unless the dendrimers are used as reagents with linear polymers or oligomers to prepare hyperbranched grafts.

2
Hyperbranched Poly(acrylic Acid) Grafts

The synthesis of hyperbranched grafts of poly(acrylic acid) (PAA) using a "graft-on-a-graft" strategy is a general method for modifying a variety of surfaces. It requires as a starting material a surface that contains some functional groups though the amplification of functionality inherent in the chemistry means that a surface with only a modest level of functional groups can produce an interface with a macroscopically detectable concentration of functional groups. Examples of surfaces that have been modified include silicon (using the hydroxyl groups of the $Si(OH)_x$ layer), gold with functional self-assembled monolayers, glass, and surface-oxidized polyolefin films and powders. In each case, robust ultrathin supported-films are the products. This covalent multistep strategy is based on functional group protection/deprotection and affords modest control over the product film thickness. In PAA grafting, this control is based on the number grafting stages that are used. The product hyperbranched grafts range in thickness from ca. 30 Å to greater than 1000 Å. The film thickness initially increases rapidly in a non-linear fashion since each additional layer is added in a branching fashion multiplying the number of grafting sites (Fig. 2). After several grafting stages the thickness increases in a linear fashion. This variable extent of progress of this grafting chemistry as measured by either ellipsometry on reflective metal surfaces or as measured by titration of the – CO_2H groups being introduced on higher surface area materials is very similar substrate to substrate (Fig. 2) [18, 19].

   The graft-on-a-graft strategy was conceived of as a synthetically "forgiving" alternative to an attempted but ineffective borane-based radical graft polymerization onto vinyl terminated self-assembled monolayers [20] on gold and was based on earlier observations that a poly(acrylic acid) graft modified with new graft sites could be used to prepare a more dense and presumably thicker graft with subsequent polymerization or grafting steps [21]. It was also conceptually more attractive than other approaches that used monomers as grafting agents because the multiple grafting of oligomeric grafting reagents could compensate for inefficiencies in reactions at surfaces as discussed above (Fig. 1).

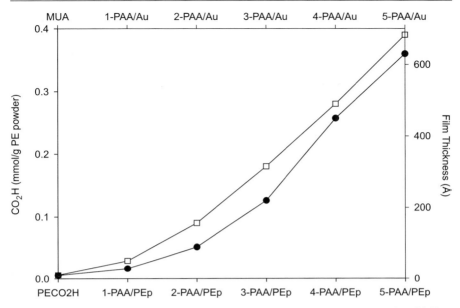

Fig. 2 Progress of hyperbranched poly(acrylic acid) graft formation on smooth gold films as measured by ellipsometry (●) or on polyethylene powders as measured by titration (□) of the supported $-CO_2H$ groups

2.1
Hyperbranched Poly(acrylic Acid) Graft Synthesis on Gold Surfaces

The synthesis of surface grafted hyperbranched films of poly(acrylic acid) was first described on gold substrates [18]. This synthesis of hyperbranched grafts of poly(acrylic acid) (PAA) on gold, shown in Scheme 1, began with a self-assembled monolayer of mercaptoundecanoic acid (MUA). Activation of the carboxylic acid groups of this monolayer was accomplished by formation of mixed anhydrides with ethyl chloroformate. While other activating agents (e.g. carbonyl diimidazole or DCC worked), the best yields were obtained with alkyl chloroformates. Subsequent amidation of this electrophilic surface by an oligomeric reagent, α,ω-diamino-poly(*tert*-butyl acrylate) (PTBA), yielded a 1-PTBA graft on MUA functionalized gold (1-PTBA/Au). This 1-PTBA/Au graft was initially converted to a 1-PAA/Au graft by acidolysis with *p*-toluene sulfonic acid/H_2O. Subsequent work showed that this acidolysis proceeded equally well using methanesulfonic acid (15 min, room temperature). Activation of the carboxylic acid groups of this first 1-PAA/Au graft with more ethyl chloroformate followed by treatment of the new polyanhydride surface with more α,ω-diamino-poly(*tert*-butyl acrylate) oligomer produced a 2-PTBA/Au graft. Acidolysis of this second graft layer of PTBA produces a 2-PAA/Au graft. Repeating

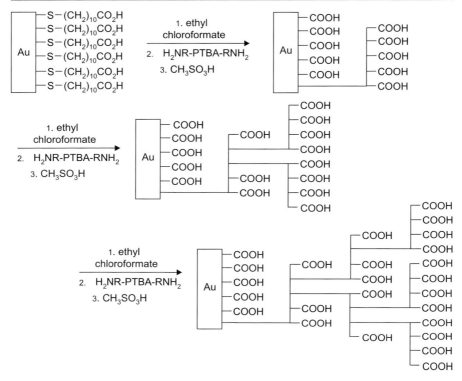

Scheme 1 Repetitive step-by-step synthetic scheme leading to formation of a hyperbranched graft of poly(acrylic acid) on a mercaptoundecanoic acid self-assembled monolayer on a supported gold film

this process for several generations produces a dense, highly functionalized surface. Films containing as many as seven graft layers were successfully prepared.

The use of appropriate functionalized oligomers is a key to the success of this synthesis. The necessary functionalized oligomers were prepared by polymerization of *tert*-butyl acrylate (n = ca. 120) with a functional AIBN initiator (Eq. 1). Since

(1)

tert-butyl acrylate polymerizations terminate mainly by coupling [22], the product is principally a difunctional poly(*tert*-butyl acrylate). The carboxylic acid terminated oligomers so-formed were subsequently converted into primary amines (Eq. 2). The product polymer was characterized by end group analysis at both the $-CO_2H$ stage and at the $-NH_2$ stage and had a M_n of 12 000–18 000 Daltons in various preparations.

$$(2)$$

    The progress of this surface graft chemistry was followed by Fourier transform infrared external reflection spectroscopy (FTIR-ERS), water contact angle goniometry, X-ray photoelectron spectroscopy (XPS), and ellipsometric analysis. The thickness of these films increased in a nonlinear fashion (Fig. 1). Ellipsometric analysis showed the thickness changed from ca. 30 Å for a single PAA/Au graft to greater than 1000 Å for films that were prepared using more than 5 grafting stages [23]. Activation of the MUA/Au film by formation of the mixed anhydride was shown to be quantitative by FTIR-ERS spectroscopy. Upon treatment with the diamino-oligomer of PTBA ($M_n = 14\,600$), evidence for the formation of a 1-PTBA/Au graft was seen in the FTIR-ERS spectrum which showed a small amide peak and a large *tert*-butyl ester peak. After acidolysis, the absorption peaks for *tert*-butyl esters disappeared. Water contact angle goniometry showed that the $-CO_2H$-rich surface was more hydrophilic as expected. The carbonyl peak intensity in the infrared spectrum increased with each additional grafting stage. The amide peaks in the infrared spectrum from covalent grafting were also detectable in the IR spectrum. These grafts were found to be stable to extensive solvent treatment. No change in the carbonyl intensity in the FTIR-ERS spectrum was observed after Soxhlet extraction with methylene chloride or sonication with acetone or acid.

    Tapping-mode atomic force microscopy studies showed that as these hyperbranched PAA films became somewhat less smooth as they increased in thickness through successive grafting stages [24]. For example, a very smooth initial single-crystal Au(111) surface with a root mean square (RMS) roughness of 0.2 nm (over a 2 µm × 2 µm area) had its roughness increased to

Table 1 Root mean square roughness of PAA grafts on Au/Si measured over a $5 \times 5\,\mu m$ area

Graft	RMS (nm)
MUA/Au/Si	2.01
1-PAA/Au/Si	1.44
2-PAA/Au/Si	1.15
3-PAA/Au/Si	1.02
4-PAA/Au/Si	1.53

0.3 nm for a 1-PAA/Au graft and 0.8 nm for a 3-PAA/Au graft. However, grafting on a rough Au/Ti/Si surface that was prepared from Au deposition on Ti/Si resulted in some surface smoothing after several grafting layers (Table 1). In this case the 1-PAA/Au, 2-PAA/Au, 3-PAA/Au, and 4-PAA/Au grafts were all smoother than the initial Au/Ti/Si surface. The PTBA grafts were generally less smooth. The relative smoothness of these surfaces leveled off after 2 or 3 graft stages and the surface became increasingly rough with additional grafting layers following the trend noted earlier with grafting of PAA on single-crystal Au(111) surfaces.

2.1.1
Derivatives of Hyperbranched Poly(acrylic Acid) Grafts

Hyperbranched grafts of poly(acrylic acid) on gold can easily be derivatized. The most common approach has been to use derivatizing agents that contain reactive amines or alcohols to form carboxylic acid amides or esters. Examples of compounds covalently incorporated into these PAA/Au interfaces are shown in Fig. 3. Equation 3 illustrates this general method which involves first activating the poly(acrylic acid) grafts with ethyl chloroformate. Subsequent treatment (here with an amine) then produces a mixture of derivatized $-CO_2H$ groups and unmodified $-CO_2H$ groups in the interface. The covalent amidation strategy has been used to prepare low-energy fluorinated surfaces [25–27]. Amidation or esterification has been used to incorporate pyrene fluorescence probes, molecular recognition elements like crown ethers cyclodextrin [28], ferrocene, poly(ethylene glycol), and dye chromophores [23]. Other chemistry typical of $-CO_2H$ groups can also be carried out with the $-CO_2H$ groups in these interfaces. This includes acid-base chemistry, reductions, ion exchange, and non-covalent modifications through hydrogen-bonding. Finally, the polyvalent nature of these films can be used to advantage in molecular assembly procedures. Specifically, these films have been noncovalently functionalized using ionic entrapment of polycations [29] and enzymes [30] and through polyvalent hydrogen bonding interactions with polyvalent hydrogen-bond acceptors [31].

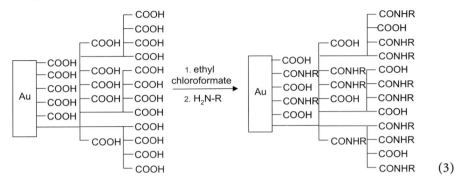

(3)

Hydrophobic, fluorinated hyperbranched grafts were synthesized from an ethyl chloroformate activated hyperbranched 3-PAA/Au graft using the fluorinated alkyl amine $H_2N(CH_2)(CF_2)_6CF_3$. Reaction of an activated 3-PAA/Au film with $H_2N(CH_2)(CF_2)_6CF_3$ produced a fluorinated film that almost doubled in thickness. XPS analysis showed 46 atom-% F in the product fluoramidated graft which is 86% of the theoretical atomic concentration for a homogenous

Fig. 3 Functionality that has been incorporated into PAA/Au interfaces through amidation or esterification chemistry

fluorinated graft. A second activation and $H_2N(CH_2)(CF_2)_6CF_3$ treatment increased the films' F content to 50 atom-%, 93% of the theoretical concentration for a homogenous graft [25]. No Au peak was observed from the underlying gold support indicating good coverage. Films were also synthesized where the internal grafts layers were fluorinated and the external grafts layers were either hydrophilic PAA grafts or fluorinated [26]. The control of the fluorine content of the core and shell of these grafts demonstrates the flexibility of this "graft-on-a-graft" strategy and the ability of later graft stages to cover defects in earlier grafts. This is illustrated by steps where the small amount of residual – CO_2H groups remaining after formation of a fluorinated graft like that described above were used as the base for formation of a hydrophilic graft with a fluorinated hydrophobic interior. In this example, a film with a hydrophobic interior was first prepared by adding $H_2N(CH_2)(CF_2)_6CF_3$ to a grafting solution of diamino-PTBA in the early grafting stages. In the final fluorine-grafting step, a grafting mixture containing a 500 : 1 mol : mol ratio of $H_2N(CH_2)(CF_2)_6CF_3$ to diamino-PTBA of 500 : 1 was used. At this point, the supported hyperbranched graft film had a water contact angle of 100°. Then two additional grafting stages were carried out using only the diamino-PTBA polymer. These steps only produced hydrophilic – CO_2H on the exterior or shell of this original fluorinated graft. Formation of this hydrophilic exterior or shell was confirmed by absence of F peaks in the XPS spectrum in the final product and by the change in contact angle from 100° to 14°.

These hydrophobic, fluorinated hyperbranched films on gold passivate the underlying gold metal toward electrochemical reactions. For example, under basic conditions a fluorinated 3-PAA/Au interface had a measured charge-transfer resistance that was 40 times more than the resistance of an unfluorinated 3-PAA film under the same conditions. These fluorinated hyperbranched grafts were significantly more passivating than a simple monolayer. For example fluorinated 3-PAA/Au grafts were shown to passivate the underlying gold to a 10^4 times greater extent than a MUA coated electrode and about 10 times more than a hexadecanethiol monolayer [25]. The fluorinated 3-PAA/Au grafts' passivation was greater under basic conditions in contrast to the unfluorinated 3-PAA/Au grafts that were more passivating under acidic conditions. For example, the fluorinated grafts' had charge-transfer resistance that increased 10-fold as pH was changed from 3 to 10 while the unfluorinated 3-PAA/Au grafts' charge transfer resistance decreased 19-fold as the pH changed from 3 to 10. Presumably that reflects the fact that PAA grafts have pH-dependent swelling (vide infra) that is lacking in the fluorinated 3-PAA/Au grafts [27].

Modification of PAA/Au grafts with the amine and alcohol functionalized pyrenes **2** and **3** produced highly fluorescent films [23]. These derivatized films exhibited both monomer and excimer fluorescence. The relative amounts of monomer and excimer emission depended on the pyrene concentration used in the derivatization process. When modest concentrations (ca. 0.05 M) of pyrene derivative 2 were used in amidation, films with only excimer fluorescence emission were produced. Mixed anhydride interfaces that were treated with very dilute concentrations (ca. 10^{-6} M) of 2 in contrast produced films that exhibited monomer fluorescence. It was also possible to incorporate pyrene during grafting. For example, films prepared by treatment of 2-PAA interfaces with a (1 : 50) (mol : mol) mixture of 2 and diamino-PTBA (in step 2 of Scheme 1) followed by hydrolysis also exhibited only monomer fluorescence. This suggests that intra-interface interactions of substituents on these hyperbranched grafts are facile even though the interfaces have some crosslinking due to the difunctionality present in the grafting oligomer $NH_2 - PTBA - NH_2$.

In a typical amidation reaction, ca. 55% of carboxylic acids were converted to carboxamides. The extent of conversion varied depending on the amine nucleophile. To quantitatively convert the $- CO_2H$ groups in the hyperbranched graft into $- CONHR$ groups, the alkyl chloroformate activation followed by amine treatment process was simply repeated. In some cases, up to six repetitions of this process were required to achieve very high conversions of acid to amide. This need for repetitive amidations reduced the synthetic utility of this process. This amidation reaction is presumably limited in effectiveness because the nucleophilic amine can attack either of the anhydride carbonyl derived from the $- CO_2H$ or the chloroformate. The former reaction leads to the desired amide product. The latter reaction produces a soluble amide and a salt of the PAA $- CO_2H$ group and the amine nucleophile. Other activation reagents (e.g. carbonyl diimidazole or DCC) were also studied. However, ethyl chloroformate was the most effective activating agent.

Esterification of the PAA grafts could be carried out either by ethyl chloroformate activation and alcohol treatment or by simple Fischer esterification. The latter reaction proved to be very effective in spite of the harsh conditions employed. Indeed, the stability in thickness of a 3-PAA/Au surface in the presence of toluene and a sulfuric acid catalyst was notable. A direct comparison of the effectiveness of various esterification procedures showed that the use of sulfuric acid as a catalyst in excess alcohol or the use of p-toluenesulfonic acid (p-TsOH) as a catalyst in toluene were both effective in esterification (Eqs. 4–6). Either acid-catalyzed procedure (Eq. 5 or 6) produced > 90% ester formation in contrast to the 40–60% ester yield using the EtOCOCl activation shown in Eq. 4. For example, a poly(ethyl acrylate) film was formed from a 3-PAA/Au using a toluene solution containing EtOH with 0.1 M p-TsOH as a catalyst after an overnight reflux. These conditions produced a product film with < 10% remaining $- CO_2H$ groups after 12 h of reflux. Ellipsometric thickness of the hyperbranched ester film showed that the original 3-PAA film thickness of 344 Å had changed to 424 Å. A concomitant change in advancing water contact angle from 24° to 93° was consistent with esterification. This result contrasted with results for a similar esterification of a MUA monolayer on gold where there was no detectable remaining monolayer after an overnight reflux in a toluene-EtOH mixture in the presence of *p*-toluenesulfonic acid. The greater stability of the hyperbranched graft vis-à-vis the monolayer was ascribed to the light crosslinking of the hyperbranched graft by the mostly telechelic oligomeric grafting agent.

$$
\begin{array}{l}
-CO_2H \\
-CO_2H \\
-CO_2H \\
-CO_2H
\end{array}
\quad
\begin{array}{c}
\text{1. ClCOOCH}_2\text{CH}_3 \\
\textit{N}\text{-methylmorpholine} \\
\xrightarrow{\hspace{2cm}} \\
\text{2. CH}_3\text{CH}_2\text{OH} \\
\text{3. NaOH (-CO}_2\text{Na formation)}
\end{array}
\quad
\begin{array}{l}
-CO_2Na \\
-CO_2CH_2CH_3 \\
-CO_2Na \\
-CO_2CH_2CH_3
\end{array}
\tag{4}
$$

$$
\begin{array}{l}
-CO_2H \\
-CO_2H \\
-CO_2H
\end{array}
\quad
\begin{array}{c}
\text{1. CH}_3\text{OH, H}_2\text{SO}_4 \\
\text{6 h reflux} \\
\xrightarrow{\hspace{2cm}} \\
\text{2. NaOH (-CO}_2\text{Na formation)}
\end{array}
\quad
\begin{array}{l}
-CO_2CH_3 \\
-CO_2CH_3 \\
-CO_2CH_3
\end{array}
\tag{5}
$$

$$
\begin{array}{l}
-CO_2H \\
-CO_2H \\
-CO_2H
\end{array}
\quad
\begin{array}{c}
\text{1. H}_2\text{C=CHCH}_2\text{OH} \\
\text{0.1 M } p\text{-TsOH, toluene} \\
\text{12 h reflux} \\
\xrightarrow{\hspace{2cm}} \\
\text{2. NaOH (-CO}_2\text{Na formation)}
\end{array}
\quad
\begin{array}{l}
-CO_2CH_2CH=CH_2 \\
-CO_2CH_2CH=CH_2 \\
-CO_2CH_2CH=CH_2
\end{array}
\tag{6}
$$

Reduction of the $-CO_2H$ groups in a PAA hyperbranched graft was of interest as a route to poly(allyl alcohol) hyperbranched grafts. Surprisingly, reducing agents known to be effective in solution-state organic chemistry for reduction of $-CO_2H$ groups to $-CH_2OH$ groups were not effective in derivatizing these PAA grafts. Reactions using BH_3-SMe_2 or $LiAlH_4$ both produced interfaces that contained unreacted $-CO_2H$ groups based on IR analysis of films after attempted reduction followed by acidification. Complete reductions (Eq. 7) were only possible if the PAA hyperbranched graft was first activated with ethyl chloroformate and then treated with BH_3-SMe_2. These poly(allyl alcohol) grafts on polyethylene substrates were subsequently used as substrates for further radical grafting using Ce(IV) [32].

$$
\text{(PAA graft with -CO}_2\text{H groups)}
\quad
\begin{array}{c}
\text{1. ClCO}_2\text{Et} \\
\xrightarrow{\hspace{2cm}} \\
\text{2. BH}_3\text{-S(CH}_3)_2
\end{array}
\quad
\text{(graft with -CH}_2\text{OH groups)}
\tag{7}
$$

Hydroxyl-functionalized hyperbranched grafts were also accessible from ester grafts by organometallic chemistry. IR analysis of the reaction of an esterified 3-PAA/Au hyperbranched graft with methylmagnesium bromide showed that the absorbance due to the ester carbonyl group completely disappeared. The thickness change (370 Å to 285 Å) and the change in the IR spectrum for this film were interpreted in terms of the simple Grignard chemistry shown in Eq. 8.

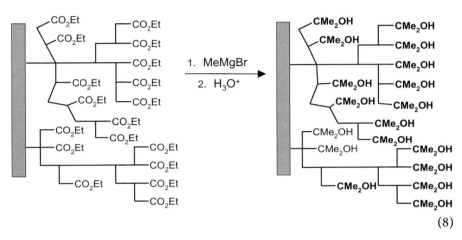

$$(8)$$

The simplest derivatization chemistry is deprotonation of the $-CO_2H$ groups of a PAA hyperbranched graft. Experiments showed that deprotonation/protonation chemistry was fast with aqueous acids and bases. The extent of protonation and deprotonation could be followed by assaying the change in absorbance of the $-CO_2H$ carbonyl (as the $-CO_2H$ group became a $-CO_2Na$ group the $\nu_{C=O}$ peak shifted to 1560 cm^{-1}) (Fig. 4). The p$K_{1/2}$ for the point where the integrated intensity of the ν_{CO_2H}/ν_{CO_2Na} absorbances was 1:1 was 4.3, a value very similar to the pK_a of poly(acrylic acid) in solution. This similarity in acid-base chemistry for these interfaces versus solution state chemistry of a similar weak acid contrasts with the very dissimilar pK_a of $-CO_2H$ groups at the surface of oxidized polyethylene versus the solution pK_a of acetic acid [33].

As noted below, such deprotonation/protonation chemistry occurs repeatedly without any loss of the hyperbranched graft as measured by ellipsometry (vide infra). This stability to acid and base mirrors the stability noted above for these hyperbranched grafts in continuous extractions and acid-catalyzed esterification chemistry and is thought to be a consequence of the light crosslinking that occurs during the grafting process.

The ready formation of polyanions from poly(acrylic acid) grafts or of polycations from a polyaminated surface leads to a polyvalent anionic or cationic interface. Such interfaces can be modified ionically (as in ionic

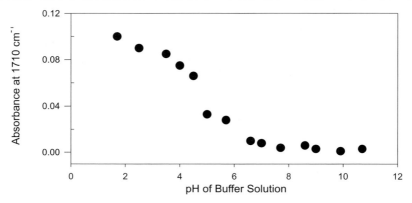

Fig. 4 The change in the absorbance at $1710\,\text{cm}^{-1}$ as a 3-PAA/Au film is immersed in buffers of varying pH. An increase in intensity of a peak assigned to $-\text{CO}_2\text{Na}$ was seen concomitant with the decrease in intensity of the $-\text{CO}_2\text{H}$ peak at $1735\,\text{cm}^{-1}$

layer-by-layer assembly) [2]. This ionic modification offers an alternative to covalent modification procedures. Electrostatic interactions between polycationic polymers or enzymes with cationic surface groups with poly(sodium acrylate) grafts were effectively used as examples of this method for post-graft functionalization of a hyperbranched interface. Linear cationic poly(allyl amine), poly-D-lysine, and dendritic PAMAM all chemisorbed into a 3-poly(sodium acrylate)/Au film. The ionically assembled composites so formed were stable to washing with polar solvents. Thickness measurements of the ionically modified hyperbranched grafts showed a significant increase due to the exchange of sodium cations with larger polymeric cations. In general, these ionic assemblies fully disassemble after acid treatment [29].

The poly(sodium acrylate) hyperbranched graft formed in a simple deprotonation was used to ionically immobilize enzymes. Studies showed that immobilized glucose oxidase in a 3-poly(sodium acrylate)/Au film retained significant activity on immobilization and that the enzyme in this ionic assembly remained active for more than 60 days of storage [30].

Derivatives of hyperbranched 3-PAA/Au grafts were also used as polymeric filters for chemical sensors [28]. This chemistry took advantage of the fact that these hyperbranched grafts are ultrathin. Earlier work that used chemically selective supported polymer membranes generally used thicker coatings which slowed mass and electron transfer. As a proof of concept, a 3-PAA hyperbranched graft was selectively modified first with β-cyclodextrin (after ClCO_2Et activation of the PAA $-\text{CO}_2\text{H}$ groups). Then this graft that contained a homogeneous dispersion of covalent immobilized β-cyclodextrin was allowed to react with N-hydroxysuccinimide in the presence of a water soluble carbodiimide (EDC). The NHS ester-containing film so formed was next allowed to react for 18 h with an aqueous solution of poly-D-

lysine in a pH 8 buffer. This produced films with a PAA-(β-cyclodextrin) bulk and a poly-D-lysine modified surface presumably because the poly-D-lysine diffusion into the bulk of the ca. 250-nm thick film was slow relative to the rate of reaction of the lysine ε-amino groups with the NHS esters.

Films formed in this way exhibited pH dependent permeability to redox probe molecules. For example, benzyl viologen did not permeate the film at low pH where the poly-D-lysine cap was protonated. However at high pH where the cap is neutral and the underlying $-CO_2H$ groups are in their carboxylate form, benzyl viologen was electrochemically reduced. Opposite pH dependence was seen with a negatively charged redox molecule, anthraquinone-2-sulfonate.

2.1.2
Aqueous Solvation of Hyperbranched Poly(acrylic Acid) Films

Hyperbranched thin films of poly(acrylic acid) can be highly solvated and their solvation is pH dependent (Eq. 9). The solvation properties of hyperbranched grafts of poly(acrylic acid) prepared on a gold substrates were studied after acid and base treatment both as "dry" films (films that had been removed from a buffer, EtOH washed and dried under N_2) and in situ in the presence of an aqueous buffer [29]. Ellipsometric analysis showed that a dry 3-PAA/Au film had a thickness of 230 Å. This thickness changed in a fully reversible manner to ca. 450 Å after deprotonation (Fig. 5). The thickness of these films was even greater in a buffer solution as measured by in situ ellipsometry. After immersion in a pH 1.7 buffer solution, the solvated acid form of the 3 PAA/gold graft swelled to a thickness of 430 Å. A buffer with pH 7.2 increased the swelling of the graft layer to 560 Å. A thickness titration curve between pH 2.6 to 7.2 for hyperbranched poly(acrylic acid) films on gold was made by in situ ellipsometric analysis. As noted above, the $pK_{1/2}$ (buffer pH at thickness midpoint) is pH 4.3 which is similar to the pK_a of poly(acrylic acid).

$$(9)$$

The pH responsiveness of a 3-PAA/Au graft that was amidated with N,N-dimethylaminoethylenediamine was also studied by in situ ellipsometry [32]. This amine-containing graft had pH dependent swelling characteristics that were opposite to those of the $-CO_2H$-rich graft because the amine-rich film

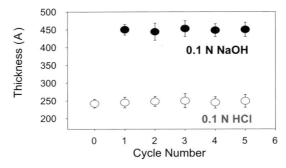

Fig. 5 Reversibility of protonation/deprotonation of a 3-PAA/Au film with 0.1 N NaOH and 0.1 N HCl respectively as measured by ellipsometry of the acidic or basic form of a 3 PAA/Au graft after removal form the acid or base solution, an ethanol rinse and drying under N_2

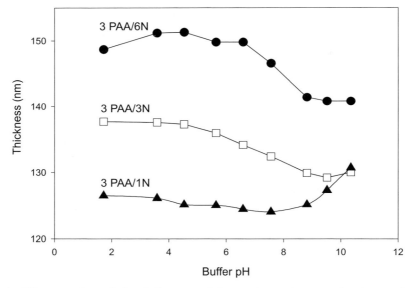

Fig. 6 Ellipsometric analysis of films containing various mixtures of amines and carboxylic acids: 3-PAA/1N (ca. 55% – CO_2H); 3-PAA/3N (ca. 17% – CO_2H); and 3-PAA/6N (no – CO_2H groups could be distinguished from carboxamide carbonyl peak in the FTIR-ERS spectrum). The 1N, 3N and 6N refers to the number of times the amidation in Eq. 3 was repeated using $NH_2CH_2CH_2N(CH_3)$ as the amidating reagent

is cationic at low pH and neutral at high pH (Fig. 6). Thickness changes measured ellipsometrically as a function of pH were also very sensitive to the extent of conversion of carboxylic acid to amide. Only after six repetitions of activation followed by amidation was the expected ellipsometric thickness change as a function of pH observed. In this case, the thickness of the basic graft decreased as the pH increased. The inflection point for this titration of thickness change was at ca. pH 7.9. Studies of pH-dependent thickness changes for films with lower conversions of acid to amide were more complicated. Such amphoteric interfaces contained variable mixtures of amine and acid groups and did not show a clear inflection point.

2.1.3
Patterning of Hyperbranched Poly(acrylic Acid)-Derived Grafts

Crooks' group has used photolithographic and μ-contact printing technology to form patterned hyperbranched grafts. This topic is the subject of a review [34] but a few examples will be mentioned here. Using photolithography, features as small as 5 μm have been prepared as shown in Scheme 2 [35]. Beginning with a 3-PTBA/Au graft, a coating of the photoacid triphenylsulfonium hexafluoroantimonate was applied to the film. Irradiating the coated 3-PTBA/Au film through a photomask with UV light caused the exposed regions of poly(*tert*-butyl acrylate) to undergo acidolysis to form poly(acrylic acid). The hyperbranched poly(acrylic acid) regions were activated with ethyl chloroformate and treated with an amine-terminated dansyl fluorophore. The patterned region was then visualized by fluorescence microscopy with an

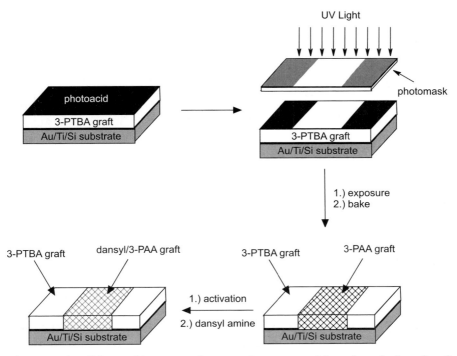

Scheme 2 Photolithographic process for preparing patterned hyperbranched grafts of poly(acrylic acid)

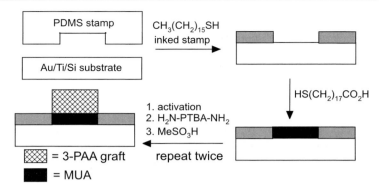

Scheme 3 Procedure used for preparing patterned hyperbranched poly(acrylic acid) films by μ-contact printing

excitation wavelength of 330–380 nm and emmision wavelength at 420 nm. A second coating of photoacid followed by UV light exposure converted the remaining regions of 3-PTBA/Au grafts to 3-PAA/Au grafts. Activation followed by treatment with the amine-terminated eosin fluorophore with an excitation wavelength of 540–580 nm and an emmision wavelength of 600–660 nm produced a patterned film with two dyes. Fluorescence microscopy showed that the dyes were segregated from each other.

Several approaches and applications of patterned hyperbranched grafts of poly(acrylic acid) by μ-contact printing have been described. An example of patterning hyperbrached grafts of poly(acrylic acid) by μ-contact printing is shown in Scheme 3 [36]. A pattern of $CH_3(CH_2)_{15}SH$ was first applied to an gold surface using a poly(dimethylsiloxane) (PDMS) stamp. The remaining naked gold regions were then functionalized with MUA and a 3-PAA/Au graft was prepared on the MUA modified gold as previously described in Scheme 1. This patterning chemistry has been extended to include polyethylene substrates [37], biopatterning [38–40], and dendrimer/poly(maleic anhydride)-c-poly(methyl vinyl ether) (Gantrez) composites [41].

2.2
Hyperbranched Grafts on Polymer Surfaces

The modification of polymeric surfaces is of great interest as it affords a way to enhance the surface properties of a polymer without affecting the bulk properties of a polymer. Thus, it is not surprising that the synthesis of hyperbranched PAA grafts has been extended to include both soft two dimensional (films, wafers, blocks) and three dimensional (powder) polymer surfaces. These approaches used the same general synthetic grafting scheme previously described for graft-on-a-graft hyperbranched grafting on hard metal surfaces.

2.2.1
Synthesis of Hyperbranched Poly(acrylic Acid) Grafts on Polyethylene Films

The surface modification of polyethylene (PE) films, PE powders, and polypropylene (PP) wafers with hyperbranched poly(acrylic acid) grafts was carried out as shown in Scheme 4 [19, 42, 43]. First, the polyolefin substrates were functionalized by oxidation with $CrO_3 - H_2SO_4$. Subsequent activation with ethyl chloroformate followed by grafting with α,ω-diamino-poly(*tert*-butyl acrylate) oligomer then produced a 1-PTBA/PE graft. Acidolysis of the *tert*-butyl esters of the 1-PTBA/PE film produced the poly(acrylic acid) or 1-PAA/PE graft. A $(1 + x)$-PAA/PE graft was prepared by repeating the ac-

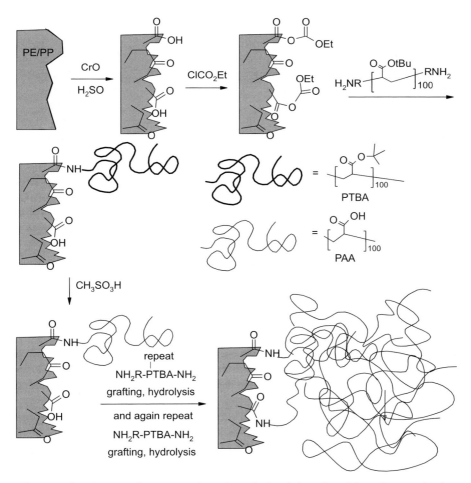

Scheme 4 Chemistry used to prepare hyperbranched poly(acrylic acid) grafts on polyethylene film, polyethylene powder or polypropylene wafers

tivation step, the α,ω-diamino-poly(*tert*-butyl acrylate) grafting, and then acidolysis x times.

The initial oxidation product of a polyethylene film or powder is a surface that includes some $- CO_2H$ groups. These groups' presence can be detected by attenuated total reflectance infrared (ATR-IR) spectroscopy as a peak at 1710 cm^{-1}. These $- CO_2H$ groups served the same purpose as the $- CO_2H$ groups of MUA/Au, serving as starting points for the hyperbranched grafting scheme. The hyperbranched grafts that were formed were characterized by water contact angle goniometry, ATR-IR spectroscopy, and XPS analysis. With each additional grafting stage the intensity of the acid carbonyl at 1710 cm^{-1} increased relative to the intensity for the C-H absorption in the ATR-IR spectrum. The poly(acrylic acid) peak at 1710 cm^{-1} was the dominant peak in the ATR-IR spectrum after 3–4 grafting stages on PE film. The carboxamide peak from the grafting was also visible at 1650 cm^{-1}.

The stability of these hyperbranched films was tested with repeated acid/base treatments and by Soxhlet extraction of grafted films with CH_2Cl_2. The reversibility of the deprotonation/protonations effected by base/acid treatment was confirmed both by water contact angle analysis and ATR-IR spectroscopy. The ATR-IR spectra showed that on base treatment that the carbonyl peak due to the acid carbonyl disappeared concomitant with the appearance of a new carbonyl peak due to formation of a carboxylate group. These spectral changes were reversible on dilute acid treatment with the carboxylic acid carbonyl peak reappearing. The relative peak intensities of IR peaks for the graft functionality in the ATR-IR spectrum of the 3-PAA/PE film after this deprotonation/protonation cycle were unchanged after repeated deprotonation/protonation cycles. Water contact angle analysis confirmed this reversibility – a protonated films' water contact angle Θ_a was 48° and the deprotonated film's water contact angle was 15°. This reversibility was consistent with that seen for similar hyperbranched grafts on gold (Fig. 5).

The uniformity of these hyperbranched PAA grafts on PE films was studied by forming derivatives with visible and fluorescent dyes. These hyperbranched films were modified by treating an activated 3-PAA/PE film with an amine terminated *p*-methyl red dye (MR) **9** or with a dansyl amine **10** to form MR/3-PAA/PE or dansyl/3-PAA/PE films, respectively. Visual inspection under a light or fluorescence microscope showed that the methyl red-labeled and fluorescently labeled films were uniform in color and fluorescence. The methyl red/3-PAA/PE film derivative showed a responsive color change with change in pH. The MR labeled 3-PAA/PE film was red after immersion in an acidic ethanol solution and yellow after treatment with a basic ethanol solution.

Since these films only contained common elements (i.e. C, O, and N), derivatives of the hyperbranched films were prepared that were more useful in XPS analysis. For example, a 3-PAA/PE film was analyzed and shown to have 67.2 atom-% C, 31.3 atom-% O, and 1.5 atom-% N by XPS analysis.

After treating the activated film with pentadecafluorooctylamine, the film contained 42.9 atom-% C, 8.9 atom-% O, 2.1 atom-% N, and 45.9 atom-% F. The fluorinated film had an advancing water contact angle of 135°, a stark contrast to the 48° water contact angle for the starting 3-PAA/PE film. Carboxylate salts formed by base treatment were also used as probes in XPS analysis [44]. For example, a 3-PAA-PE film was treated with an ethanolic CsOH solution that produced a film that contained 6.3 atom-% Cs.

2.2.1.1
Polyvalent Functionalization of Supported Hyperbranched Films

The hyperbranched poly(acrylic acid) graft films' – CO_2H-rich interface on polyethylene can be modified by noncovalent methods just like CO_2H-rich interfaces of PAA/Au grafts. This was shown by treating deprotonated 3-PAA/PE films with cationic polyelectrolytes like poly-D-lysine, and amine terminated PAMAM dendrimers at pH 7 [31]. Equation 10 illustrates the entrapment of PAMAM dendrimers in a 3-poly(sodium acrylate)/PE film. In these cases, polyvalent entrapment of the cationic electrolyte was evidenced in the ATR-IR spectrum by the appearance of amide $C = O$ and $N – H$ peaks of the guest dendrimer that were not present in the host 3-poly(sodium acrylate)/PE film.

(10)

Analysis by XPS after incorporation of the linear or dendritic cationic polymers showed an increase in the atom % N consistent with incorporation of an N-rich polycation. The polyvalent interactions between the cationic polyelectrolytes and the poly(sodium acrylate)graftson PE were stable to Soxhlet extraction with 95% ethanol, repeated acid/base treatment, and soaking or sonication in isopropanol and THF. However, in these cases, simple acid treatment did not release the cationic polymer as was the case with PAA/Au grafts even though grafting in both cases was reportedly ionic and not covalent.

Both poly(acrylic acid)/PE and poly(acrylamide)/PE hyperbranched grafts (derived from poly(acrylic acid)/PE using ethyl chloroformate and H_2NR) can react with soluble polyacrylamide or poly(acrylic acid) reagents by hydrogen bonding. This was also shown to be a viable noncovalent method for modification of these hyperbranched grafts and constitutes a mild method for further grafting (Scheme 5) [31, 42]. In this chemistry, polymers made of hydrogen bond donors (poly(acrylic acid)) or acceptors (polyacrylamide) self assemble into the complimentary donor or acceptor hyperbranched graft interface. While a hydrogen bond is normally considered a weak bond, polyvalency magnifies the importance of each individual intermolecular hydrogen bond to produce in the aggregate strong bonding that forms a tenacious composite of the original hyperbranched graft and the added polymer [45]. This chemistry provided a readily reversible method for further grafting at a surface. For example, immersing a hydrogen-bond donating 3-PAA/PE

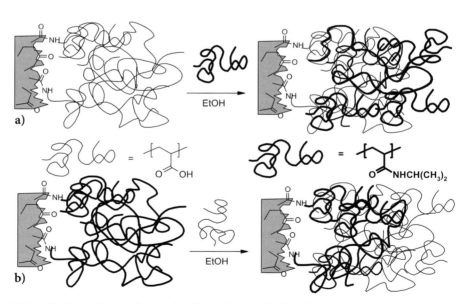

Scheme 5 Polyvalent hydrogen bonding of **a** a soluble poly(N-isopropylacrylamide) to a 3-PAA/PE hyperbranched graft or **b** a soluble poly(acrylic acid) to a hyperbranched poly(N-isopropylacrylamide) graft on PE

film into a hydrogen bond accepting poly(*N*-isopropylacrylamide) (PNIPAM) polymer solution formed a new hydrogen bond graft. This grafting was characterized by ATR-IR spectroscopy and XPS analysis. The amide peaks from the hydrogen-bond grafted PNIPAM increased in the ATR-IR spectrum of the film, and the atom concentration of N increased from ca. 1.5 atom-% N for a 3-PAA/PE film to ca. 5.8 atom-% N for the hydrogen-bond grafted 3-PAA/PE film. In these experiments, the initial 3-PAA/PE film contained a small amount of nitrogen because the film was assembled from amine-terminated poly(*tert*-butyl acrylate) oligomers and a surface-bound mixed anhydride (cf. Schemes 1 and 4).

The hyperbranched hydrogen-bond grafts formed in these self assembly processes are stable to solvent extraction. This was shown by preparing fluorescently labeled poly(*N*-isopropyl acrylamide) **11** and poly(acrylic acid) **12**.

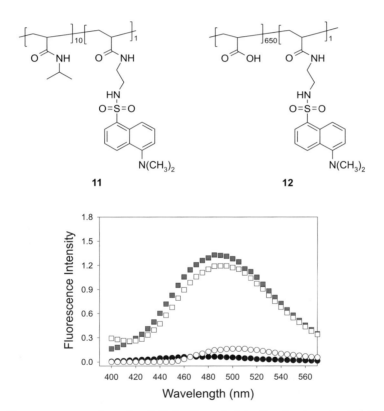

Fig. 7 Fluorescence spectra of a 5-PAA/PE graft (a) after treatment with **10** (*filled squares*); (b) after washing with THF in a Soxhlet apparatus for 24 h (*open squares*); (c) after treatment with NaOH (*open circles*). For comparison purposes, a fluorescence spectrum for an oxidized polyethylene film without any hyperbranched graft (but with – CO$_2$H groups) that was also treated with **10** is also shown (*filled circles*)

Either fluorescently labeled polymer could self assemble into a 3-PAA/PE graft. Only the hydrogen-bond donor polymer **12** assembled into a poly(*N*-isopropylacrylamide) hyperbranched graft (prepared from 3-PAA/PE using the chemistry in Eq. 3). The fluorescent films so formed had fluorescence that was much greater in than that seen in treatment of a simple oxidized PE film with **11** or **12** (Fig. 7). Moreover, the fluorescence intensity of a hydrogen-bond grafted 3-PAA/PE film showed essentially no change after 45 h of Soxhlet extraction with THF. Reaction of the hydrogen-bond grafted 3-PAA/PE film with basic ethanol followed by THF washing led to disassembly of the hydrogen bonded graft.

2.2.1.2
Grafting onto Hyperbranched Grafts

Hyperbranched grafts have also been used as substrates for further covalent grafting chemistry. One method involved the reduction of the PAA film to poly(allyl alcohol)s which were then converted to trichloroacetate esters for $Mn_2(CO)_{10}$-promoted free radical grafting [46]. For example, a trichloroacetate-containing ester interface was used successfully in grafting of more PTBA onto the interface using $Mn_2(CO)_{10}$ and light initiation. In this case, the trichloroacetate-containing ester reacts with $Mn(CO)_5$ generated by photolysis of $Mn_2(CO)_{10}$ to form radicals that can react in surface-initiated grafting monomers in solution. Monomers that were successfully used include styrene (2 M in benzene), acrylonitrile (3 M in benzene) and *tert*-butyl acrylate (1 M in benzene) [32]. These hyperbranched poly(allyl alcohol) films were also used as scaffolds for Ce(IV) redox initiated free radical polymerization of acrylates and acrylamides [32]. This second example of free radical grafting used the known Ce(IV) redox-initiated grafting procedure [47] and was successfully used for free radical grafting of the water soluble monomers methacrylic acid, acrylic acid, acrylamide and *N,N*-dimethylacrylamide onto polyethylene.

Hyperbranched grafts are interfaces and functional groups within interfaces like these can interact with one another as was noted in fluorescence studies of pyrene-labeled PAA/Au grafts. Such interactions are presumably entropically disfavored as they produce a more crosslinked species. The extent to which covalent bonds can form within these functional interfaces has been studied using thiophene oligomerization [48]. Beginning with hyperbranched grafted PAA on PE films, thiophene monomers were introduced by amidation and esterification chemistry. The thiophene-derivatized grafts were then polymerized by oxidation with $FeCl_3$ (Eq. 11). The extent of oligomerization (and, indirectly, the extent to which functional groups within the interface can react with one another) was then probed spectroscopically by monitoring the characteristic emission of thiophene oligomers.

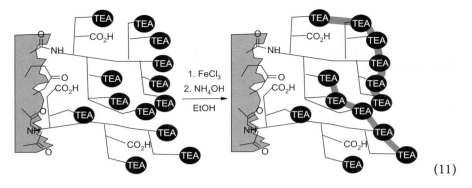

(11)

The necessary thiophene-monomer containing surface was prepared using amidation with aminothiophene derivatives or by esterification using a thiophene containing a pendant hydroxyethyl group. The amide thiophene derivative was prepared by ethyl chloroformate activation of a 3-PAA/PE interface. Subsequent reaction with 2-thiopheneethyleneamine (TEA) produced a surface that had about 50% amide based on the equal intensities of the ν_{CO_2H} and ν_{CONHAr} carbonyl peaks in the ATR-IR spectrum. The extent of amidation of the interface was increased to 65% after a second treatment. Alternatively, the 3-PAA/PE hyperbranched graft was esterified with 2-(3-thienyl)ethanol using H_2SO_4 as a catalyst. In addition to the expected appearance of amide and ester peaks in the ATR-IR spectrum of these products, an absorption at 777 cm^{-1} due to the thiophene rings was also observed confirming thiophene incorporation into the graft. The coupling yield for the esterification reaction was greater than 90%. While the yield for the esterification method is superior to the amidation chemistry, esterification involved very long reaction times (65 h) compared to 2–3 h reaction times for coupling by amidation chemistry.

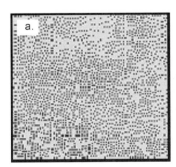
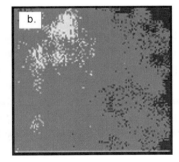

Fig. 8 Fluorescence micrograph of 0.025 mm^2 area of PE film **a** before oligothiophene formation showing no fluorescence and **b** after oligomerization show even fluorescence due to the presence of oligothiophene throughout the interface. The *figures* here are false colored *white* to show no fluorescence in (**a**) and *gray* to show fluorescence in (**b**)

Oligomerization of the immobilized thiophene monomers within the interface was accomplished using FeCl$_3$. The oligomerized film fluoresced under a UV lamp. The uniformity of the film was assayed by fluorescence microscopy (Fig. 8). While there are slight variations as shown in the fluorescence as shown in Fig. 8, analysis of a $50\,\mu m \times 50\,\mu m$ region of the oligothiophene-PAA/PE composite film showed less than 8% variation in fluorescence intensity. This uniformity in a 2-dimensional sense is consistent with the uniformity in coverage of the grafting seen in earlier fluorescence and UV-visible assays and with the relative smoothness of these hyperbranched grafts. Based on the known correlation between chain length and excitation and emission wavelengths, [49] the oligomerization reaction was estimated to couple 6 to 9 thiophene rings.

2.2.2
Hyperbranched Grafts on Polypropylene Wafers

Hyperbranched grafts of poly(acrylic acid) were prepared on polypropylene (PP) wafers using the same method as described for PE films [43]. Unlike the PE grafting chemistry, a carbonyl peak was not noticeable in the ATR-IR spectrum until the 2-PAA/PP grafting stage. This peak became a dominant feature in IR spectrum of the 5-PAA/PP graft. These hyperbranched grafts could be synthetically modified just like the PAA/PE films and were amenable to the same polyionic interactions as described for PE. For example, 5-PAA/PP had an advancing water contact angle of $61°$ that changed to $27°$ after formation of a sodium carboxylate salt. Likewise, formation of a fluoramide using ethyl chloroformate and $H_2NCH_2(CF_2)_6CF_3$ had 40 atom % F by XPS analysis and an advancing water contact angle of $124°$.

Conversion of a 5-PAA/PP hyperbranched graft into a basic graft was accomplished by amidation using $H_2NCH_2CH_2N(CH_3)_2$ after ethyl chloroformate activation. Three cycles of this activation followed by amidation were required to achieve > 90% conversion of the $-CO_2H$ groups into aminoamide groups. The resulting basic surface that formed in this chemistry also proved amenable to modification with an acidic polymer like **13** (Eq. 12). Indeed, when this $-N(CH_3)_2$ rich surface was allowed to react with the pyrene-labeled poly(acrylic acid) **13** in a pH 7 buffer solution, a fluorescently labeled PP film formed. Subsequent treatment of this surface with 0.1 N aqueous NaOH for 2 h disassembled this ionic assembly (Fig. 9).

Given that significant amounts of poly(acrylic acid) were introduced onto a polypropylene wafer's surface and given that polar polymers can penetrate into this interface it is not surprising that hyperbranched grafting affects polypropylene's mechanical adhesion strength [43]. Adhesion tests comparing the adhesive strengths of hyperbranched grafted PP and unmodified PP using a double cantilever beam test with a commercially available epoxy adhesive showed that the adhesion strength as measured by the critical strain

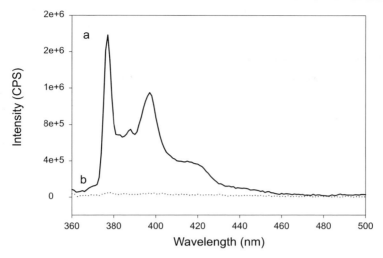

Fig. 9 Fluorescence spectra of a 5-PAA/PP wafer that was first aminoamidated with $NH_2CH_2CH_2N(CH_3)_2$ and then treated with **11** in a pH 7 buffer. The first spectrum (a) is the ionic assembly that forms at pH 7 from ion pairing of $-N(CH_3)_2H^+$ groups of the graft with $-CO_2^-$ groups of the poly(acrylic acid). The second spectrum (b) shows complete removal of **11** after treatment of the interface with 0.1 N NaOH (which converts the $-N(CH_3)_2H^+$ groups to $-N(CH_3)_2$ groups)

energy release rate of unmodified but physically roughened PP was $2\,J/m^2$ while this same value for grafted PP averaged ranged from $20–30\,J/m^2$.

$$(12)$$

2.2.3
Hyperbranched Grafts on Polyethylene Powders

The hyperbranched grafting chemistry used with polyolefin films and wafers was also extended to modification of 200 μm diameter high density PE powder substrates [19]. However, because of the relatively high surface area of the 200 μm-diameter polyethylene powder (the measured BET surface area was

$0.7 \, m^2/g$), hyperbranched grafting produced a titratable amount of CO_2H groups PAA after only a few stages of grafting. For example, while a 1-PAA/PE powder had a virtually undetectable ca. $0.02 \, mmol$ of $-CO_2H/g$ of powder, a 4-PAA/PE powder sample had $0.4 \, mmol$ of $-CO_2H/g$ of powder. Analysis by titrimetry of the increase in $[CO_2H]$/graft stage directly paralleled the results seen in thickness changes based on ellipsometric analysis for PAA hyperbranched grafts on gold (Fig. 1).

The chemistry and procedures for modification of the $-CO_2H$ groups of PAA hyperbranched grafts on PE powder were analogous to those used for PAA grafts on PE or PP films and wafers. For example, a 90% yield in ester formation was possible using acid-catalyzed Fisher esterification. Likewise, quantitative reduction (ethyl chloroformate activation, borane-dimethyl sulfide reduction) to hyperbranched poly(allyl alcohol)s and amidation all could be carried out using procedures like those used for PAA/Au surfaces.

2.2.3.1
Hyperbranched Grafts as Catalyst Supports

Polyethylene powders that have been modified with hyperbranched grafts were shown to be useful as supports for both heterogeneous and homogeneous Pd(0) catalysts. In the first example of this chemistry, a route to dispersions of Pd(0) crystallites that are active hydrogenation catalysts was described. The catalysts were synthesized by first exchanging the poly(sodium acrylate) grafts with $Pd(OAc)_2$. The Pd(II)-containing hyperbranched graft that resulted was then exposed to H_2, forming Pd(0) (Eq. 13).

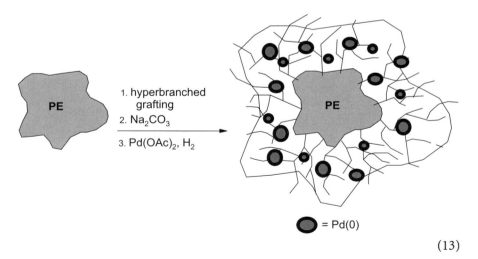

$$(13)$$

These Pd colloids dispersed in a poly(acrylic acid) interface were essentially a supported Pd/poly(acrylic acid) nanocomposite and these nanocomposite catalysts were shown to have excellent reactivity in simple hydrogenations. Limited studies showed that these catalysts were also active in other Pd(0) chemistry (e.g. in Suzuki couplings) [50]. An attractive feature of these hyperbranched graft Pd nanocomposites supported on PE powder was the physical stability of the support. For example, these PE-supported catalysts had the physical robustness associated with the bulk PE support. They did not require special procedures such as a "tea bag" to prevent the physical degradation seen with more commonly used polystyrene-supported catalysts [51, 52].

Hyperbranched grafts on PE powder were also suitable as supports for analogs of homogeneous phosphines-ligated Pd(0) catalysts (Eq. 14) [53]. In this case, a phosphine ligand was first introduced into the poly(acrylic acid) hyperbranched graft by reaction of a ethyl chloroformate activated 4-PAA/PE graft with 3-diphenylphosphinopropyl amine. The resulting 3-diphenylphosphinopropyl amide-containing polymer was then treated with Pd(dba)$_2$ to form a yellow polyethylene powder-supported phosphine-ligated Pd(0) catalyst. This catalyst was active in allylic amination chemistry and was shown to be recyclable for up to 5 cycles. As is true for homogeneous phosphines-ligated Pd(0) catalysts, exposure of the catalyst to adventitious oxygen oxidized the phosphines ligands leading to a darkening of the support due to formation of Pd colloids.

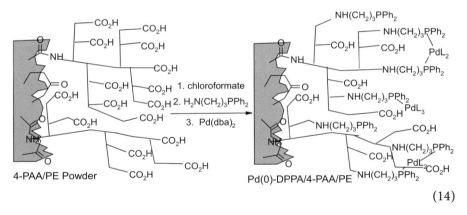

$$(14)$$

Because of the relatively high loading of functional groups on these hyperbranched PE powders, it was feasible to characterize the products and intermediates in this catalysts synthesis by ^{31}P CP-MAS NMR spectroscopy, ATR-IR spectroscopy, and XPS analysis. ^{31}P CP-MAS NMR spectroscopy was especially useful for in distinguishing the phosphinated powder, phosphine-palladium complex, and any adventitiously formed phosphine oxide. Similar NMR analyses were not successfully carried out on hyperbranched grafts on PE films. However, when this same phosphine ligand synthesis and introduction of Pd was carried out on a PE film sample, it was possible to analyze the products by XPS spectroscopy. In that case, multipoint XPS analysis over a 1 mm^2 region of a Pd(0)-DPAA/3-PAA/PE film showed that the P/Pd ratio varied only 2% from the expected 3/1 value indicating both that the complex immobilized was similar to that seen in solution and that the functionalization was homogeneous in a two dimensional sense.

3
Hyperbranched Nanocomposites

Hyperbranched multilayer composites prepared by condensation reactions of amine-containing polymers and commercially available electrophilic polymers like poly(maleic anhydride)-c-poly(methyl vinyl ether) (Gantrez) copolymers are a second type of covalent hyperbranched polyfunctional graft. These hyperbranched nanocomposite grafts have been prepared using glass, silicon, gold, aluminum, alumina and polymers as substrates [54–57]. A typical synthesis scheme leading to these sorts of ultrathin-film nanocomposites composites starting with amine-functionalized silicon using a 4th generation – NH$_2$-functionalized PAMAM dendrimer is shown in Scheme 6. In this instance, the synthesis began with an amine-rich silicon surface that was prepared by modification of the native oxide layer of silicon with aminopropyltriethoxysilane. This resulting amine-rich surface was first treated with the polyanhydride polymer Gantrez. Since an excess of Gantrez was used, this produced a new surface that contained some amic acids from reaction of surface-bound amines and anhydrides of the Gantrez copolymer along with a large excess of unreacted anhydride groups. Even though all the amines at the surface probably did not react, the surface at this stage mostly contained unreacted anhydride groups. This new anhydride-rich, electrophilic surface was then allowed to react with a dendrimer that contained nucleophilic – OH or – NH$_2$ groups on its periphery. Both PAMAM dendrimers and cascade dendrimers were successfully used in this chemistry. This reaction also used an excess of the solution-phase reagent. The resulting surface contained covalently bound dendrimers that were attached to the anhydrides via acid-ester or acid-amide (amic acid) bonds. Since an excess of the dendrimer was used, the surface was at this point rich in dendrimer and thus rich in – OH or – NH$_2$ groups. At this stage, unreacted anhydride groups from the earlier stage presumably had reacted with the reaction solvent ethanol to form acid-ester groups and were not involved further in this covalent assembly chemistry. Repetition of this sequence of reactions served to build up a covalent composite of dendrimer and the polyanhydride polymer layer by layer with covalent amic acid or ester links between the reactive components.

The progress of these reactions was followed by various techniques (Table 2). Ellipsometric analysis showed that the increase in film thickness after treating a silicon/Gantrez film with a fourth-generation PAMAM den-

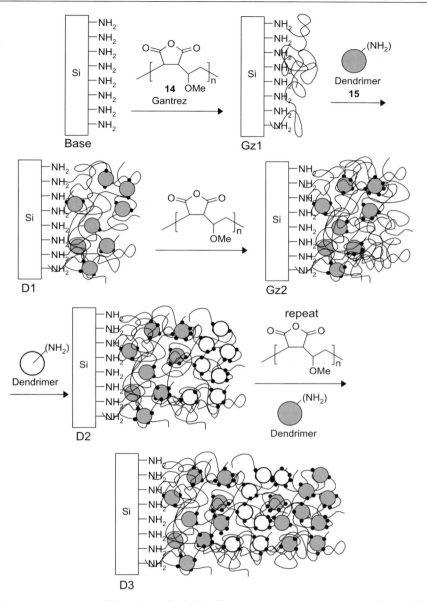

Scheme 6 Formation of hyperbranched thin film nanocomposites using an electrophilic polymeric reagent **14** and a nucleophilic amine-functionalized PAMAM dendrimer **15**

drimer was greater than the diameter of the dendrimer indicating that the dendrimer addition steps in the process incorporate more dendrimer than a single dendrimer monolayer. Similar ellipsometric analysis of the increased

Table 2 Contact angle measurements, XPS analysis, and ellipsometric thickness of Gantrez/4th Generation PAMAM dendrimer composites

	Θ_a (°)	% C	% O	% Si	% N	Thickness (nm)
Base	64	30	37	28	4.9	4.2
Gz1	70	64	31	3	0.8	6.7
D1	31	68	20	0	12.0	14.6
Gz2	56	69	30	0	1.8	19.3
D2	31	65	19	0	17.0	29.4
Gz3	55	—	—	—	—	34.9
D3	29	67	17	0	16.0	45.9
D3 (+ Δ)	97	69	16	0	14.0	39.1

thickness for the second and subsequent synthetic steps where polyanhydride polymer was added to the dendrimer-rich surface also showed that the increase in thickness for this step was greater than that seen in the addition of the first Gantrez layer to the aminated silicon surface.

The thickness of these PAMAM dendrimer/Gantrez composites is slightly dependent on the dendrimer size. For example, composites that are made with a smaller second-generation PAMAM dendrimer were shown to have an ellipsometric thickness of 33 nm at the D3 stage, compared to 46 nm for the larger fourth generation dendrimer. The molecular weight of the Gantrez used in this chemistry did not have an effect on the interface thickness.

Syntheses of similar nanocomposites on substrates such as gold, aluminum, and functionalized polymers have also been carried out with the main differences being in the first steps of the reaction. For example, in the case of gold, the initial step involved activating the $-CO_2H$ groups in a monolayer of MUA by ethyl chloroformate and treating this material with the dendrimer. Alternatively films that are composed of $-OH$ terminated dendrimers on Au substrates were used as substrates. In this case, a hyperbranched nanocomposite graft was prepared by allowing this monolayer of 11-mercaptoundecanol to react first with the polyanhydride polymer and then with the hydroxyl-terminated dendrimer. In the case of aluminum substrates, the native oxide layer at the surface was reactive enough to immobilize a thin film of Gantrez directly to the surface to produce the first layer.

The films formed in these covalent layer-by-layer assembly procedures are similar to the ionic assemblies prepared by other layer-by-layer self assembly procedures. Unlike the more defined Langmuir-Blodgett multilayer assemblies that have well defined layers, the layers of electrophilic polymer and nucleophilic polymer are intermixed to an extent. Similar effects are seen in ionic layer-by-layer assemblies.

Most applications of these nanocomposite films have focused on the effects of these films on the electrochemical reactivity of their metal supports. For example, the hyperbranched nanocomposite graft formed in Scheme 6 was used as a pH-switchable permselective supported membrane. In this case, the nanocomposite formed in this chemistry consisted of $-CO_2R$, $-CONHR'$, $-CO_2H$ and NH_2 groups. When placed in an acidic solution, this amphoteric supported membrane contained $-CO_2H$ and $-NH_3^+$ groups and was net cationic. Under these conditions, it was impermeable to electroactive cations like $Ru(NH_3)_6^{3+}$ but permeable to electroactive anions like $Fe(CN)_6^3$ (Fig. 10). Under basic conditions, this same nanocomposite film contained $-CO_2^-$ and $-NH_2$ groups and had a net anionic charge. Under these conditions, the film was only permeable to cations. At neutral conditions, this supported amphoteric membrane was permeable to both cations and anions [54]. Similarly made dendrimer/Gantrez composites prepared on high-surface-area alumina have been described by Crooks' group. These latter films on alumina substrates limit or prevent the adsorption of vapor phase volatile organic hydrocarbons to the alumina substrate with the extent of this effect depending on the film thickness [57]. This effect of hyperbranched film structure on permeability was also seen with other hyperbranched grafts as noted earlier [25, 27].

Thermal treatment of the nanocomposite film formed in Scheme 6 converts the amic acid groups of the film to imides (Eq. 15). In addition to evidence by IR spectroscopy, ellipsometry showed that the film thickness decreased by ca. 15% and contact angle goniometry showed that the film became more hydrophobic (see Table 2). This thermally treated film also did not have the pH dependent permeability of redox-active ions previously described for dendrimer/Gantrez films on gold substrates [55].

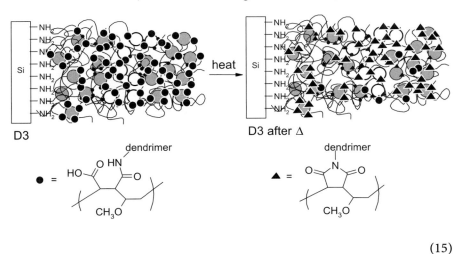

(15)

Indeed, the thermally treated film was completely impermeable to either anion. The very low permeability of these films was explained by the notion

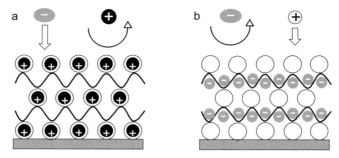

Fig. 10 pH-Dependent permeability of the anions $Fe(CN)_6^{3-}$ and the cation $Ru(NH_3)_6^{3+}$ in nanocomposite dendrimer-poly(maleic anhydride) grafts that contain $- CO_2H / - CO_2^-$ and $NH_2 / - NH_3^+$ groups in the film with **a** anion permeability at low pH where the $- NH_2$ groups in the film are protonated; and **b** cation permeability at high pH where the $- NH_2$ groups in the film are neutral and the $- CO_2^-$ groups make the film anionic

that these interfaces are thin film thermosets. Along with imide formation, multilayer composites based on PAMAM/Gantrez films are expected to undergo retro-Michael addition reactions to form free amines and acrylamides (Fig. 11). Reversal of this retro-Michael reaction in both an intra- and inter-dendrimer sense is expected to produce a highly crosslinked film. Crosslinks are presumably also formed from amines (either unreacted amines or amines

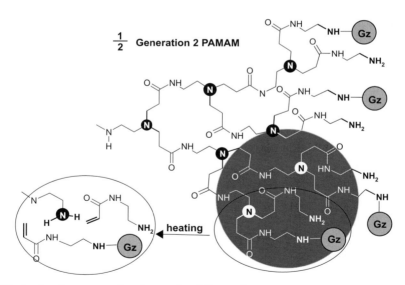

Fig. 11 In situ formation of amine nucleophiles and α,β-unsaturated carboxamides that react further in thermosetting of hyperbranched dendrimer-polyanhydride nanocomposite thin films

formed in situ) undergoing additional imide formation with half acid-half ester groups from anhydrides that did not form amic acids in earlier steps.

In work like that described in Scheme 6, Bruening reported the synthesis of multilayer composites by alternating deposition of Gantrez and the linear polyvalent nucleophile poly(allylamine) [58]. The impermeable films that were produced by this multistep hyperbranched graft chemistry were shown to passivate aluminum surfaces.

An approach that is related to these examples was used in the preparation of hyperbranced surfaces useful in conjugation of biological substrates like polynucleotides to surfaces [59]. In this chemistry (Eq. 16), an amine-functionalized glass slide was first activated by treatment with trifunctional cyanuric chloride (16). The selective reaction of the most reactive chloride [60] of the cyanuric chloride with the amine groups of the surface introduces a dichlorotriazene. One or both of the remaining chloride leaving groups of the triazene was then allowed to undergo a nucleophilic aromatic substitution reaction with hyperbranched polyethyleneimine (PEI). The resulting hyperbranched amine-rich surface that formed was then allowed to react with more cyanuric chloride. In chemistry that is conceptually like that in Scheme 6, this led to an electrophilic dichlorotriazene-rich surface with some crosslinked PEI. While this process could have been continued, at this point these surfaces were used to bind polynucleotides.

(16)

4
Hyperbranched Grafting by Surface Initiated Ring Opening Polymerization

Thus far this review has emphasized methods of producing hyperbranched grafts using graft-on-a-graft chemistry involving condensation of a carboxylic acid derivative with an amine or hydroxy nucleophile. This chemistry typically requires activation of a surface bound carboxylic acid or the use of activated carboxylic acid derivatives. In most cases, it requires the preparation of suitable functionalized polymers for grafting too, though commercially available materials are used in some instances. Alternative strategies for grafting rely on ring opening polymerizations. These strategies include ring opening polymerizations using heteroatom nucleophiles [61–65] and ring-opening metathesis polymerization (ROMP) [66–71]. Both of these synthetic methods have been used to produce linear grafts on surfaces. Hyperbranched grafts can also be produced from ring opening polymerizations using multifunctional monomers or monomers that produce polyvalent nucleophiles. Recent work by both Huck's [72] and Park's [73, 74] groups show how hyperbranched grafts can be prepared by ring opening polymerization. This approach of hyperbranched grafting from a monomer is particularly attractive since it can eliminate the repetitive steps that are typically involved in producing hyperbranched thin films.

4.1
Grafting Hyperbranched Polyglycidol

Huck's group recently explored new hyperbranched grafting syntheses using Si – OH groups as initiators for the anionic ring opening polymerization of glycidol [72]. This grafting chemistry shown in Eq. 17 begins with deprotonation of the Si – OH surface with sodium methoxide. This deprotonated surface (initiator) is then dried and immersed in the neat glycidol monomer and heated at 110 °C. The polymerization reaction begins with the anionic surface attacking the less substituted epoxide carbon. This forms an ether bond and secondary alkoxide group that can react with another monomer or generate a primary alkoxide nucleophile by proton transfer of the neighboring primary alcohol. The mechanism of growth was studied by ellipsometry. For thicker films, mulitiple cycles could be used. In subsequent cycles, the grafted polyglycidol is again deprotonated by treatment with sodium methoxide and the resulting anionic surface is immersed in more monomer. Ellipsometry showed that these films grew to 15 nm in a single reaction and that this thickness could be increased to ca. 70 nm after a third cycle. The increase in thickness from cycle to cycle increased after each grafting stage. This change in thickness in each cycle was explained by assuming that the initiator concentration (the hydroxyl groups of the grafted polyglycidol) was increasing from cycle to cycle and is similar in concept to what was seen in hyperbranched grafting via condensation polymerization (Fig. 1).

In this example of hyperbranched grafting, it was possible to study the grafted polyglycidol polymer by NMR spectroscopy. While the surface graft was not studied in situ, it was possible to characterize a graft after cleavage of the graft polymer from a surface. To accomplish this, a hyperbranched grafted material was prepared on a higher surface area support. The product polymer was then cleaved from the support and analyzed by NMR spectroscopy. For this experiment, silica gel was used as the substrate for grafting

(17)

and the polymer was cleaved from this support using HF. The recovered polymer was studied by ^{13}C NMR spectroscopy. The degree of branching (DB) was determined to be 0.31 using the peak intensities of known peaks in the ^{13}C NMR spectrum. This contrasted with the degree of branching of 0.46 for polyglycidol prepared from bulk polymerization of glycidol using 1,1,1-(trishydroxymethyl)propane as an initiator. The differences between the solution polymer and the hyperbranched graft were rationalized by assuming that the lower degree of branching for the surface initiated hyperbranched polyglycidol was due to high monomer to initiator ratios. This change was presumed to lead to higher propagation rates. The higher steric demands of a surface-initiated polymerization were also assumed to be important. The cleaved polymer was also characterized by gel permeation chromatography which showed that the product had an M_n of 5800 and polydispersity (M_w/M_n) of 1.8.

4.2
Grafting of Hyperbranched Poly(ethyleneimine)

Park's group has explored hyperbranched graft chemistry of aziridine on amine functionalized silica surfaces (Eqs. 18 and 19) [74]. Using silicon and silica substrates that have been functionalized with (3-aminopropyl)diethoxymethylsilane as initiator substrates, hyperbranched polyethyleneimine grafts were synthesized using aziridine monomers.

The growth of these films was characterized by measuring the ellipsometric thickness and measuring the concentration of primary amine groups in the grafts using an indirect spectrophotometric assay for the amount of amine groups. The method used to determine the concentration of primary amines contained within the interface was to first form an imine of surface-bound amines with 4-nitrobenzaldehyde using a surface with known surface area. The surface-bound imine was then separated from excess soluble benzaldehyde and then hydrolyzed to generate a solution of the aldehyde amines

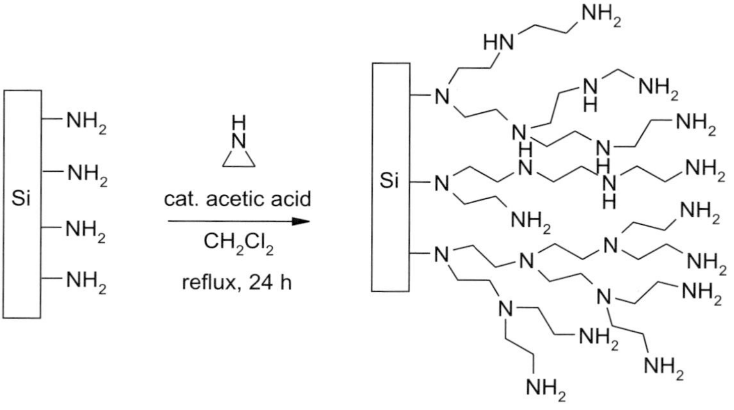

(18)

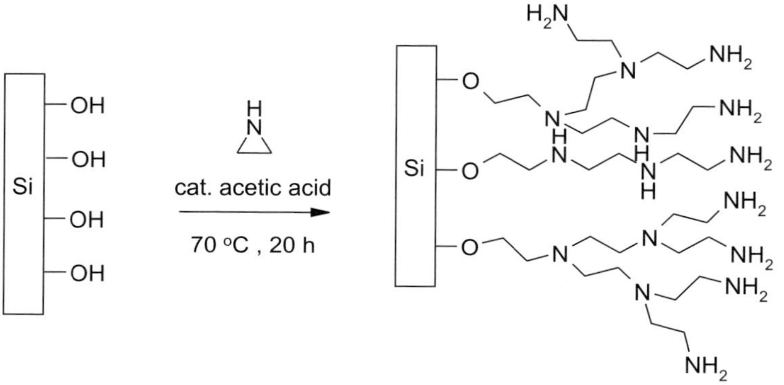

(19)

contained within the interface was to first form an imine of surface-bound amines with 4-nitrobenzaldehyde using a surface with known surface area. The surface-bound imine was then separated from excess soluble benzaldehyde and then hydrolyzed to generate a solution of the aldehyde that was analyzed by UV-visible spectroscopy [75]. A starting surface that had been functionalized with (3-aminopropyl)diethoxymethylsilane was shown to have 3.5 amines/nm^2 and a film thickness of $8\,\text{Å}$. After a typical reaction with aziridine, the film thickness increased to $36\,\text{Å}$. By assaying the change in surface density of primary amines with time, it was possible to show that this grafting was producing a hyperbranched polymeric graft. The surface density of primary amines was measured to be 66 amines/nm^2 after reaction of the amine-functionalized surface with aziridine for $24\,\text{h}$. This corresponds to a very dense graft with a concentration of 18 amines/nm^3. The maximum concentration of total amines for polyethyleneimine assuming that polyethyleneimine has a density of $1\,\text{g/cm}^3$ would be ca. 14 amine groups/nm^3. This increase in density of amine groups is in accord with formation of a hyperbranched graft and is presumed to be the result of reaction of the secondary amines of a linear graft with aziridine monomer. If the secondary amines were not reactive and if no hyperbranching had occurred, the original density of primary amine groups would have remained constant.

Park's group later reported the hyperbranched grafting of aziridine to fused silica, silicon wafers, and glass [73]. Grafting was achieved by heating the aziridine monomer solution containing a catalytic amount of acetic acid to 70 °C in the presence of a cleaned silica substrate. The thickness and primary amine density increased rapidly during the first 10 h and reach a maximum after 40 h. After grafting for 20 h the thickness of the graft layer as determined to be 27 Å and the primary amine density was reported to be 23 amines/nm^2.

5
Hyperbranched Grafts of Organic/Inorganic Hybrid Polymers

While polymers most commonly consist of carbon, nitrogen and oxygen, hybrid materials containing metals or other elements represent an important class of materials [76–78]. Hybrid materials too have been used as hyperbranched grafts. Two examples of such materials are described here. The first is a polysiloxane graft. The second is a dendritic coordination polymer based on a thermally and oxygen stable Pd(II) pincer complex [79].

5.1
Polysiloxane Hyperbranched Grafts

The synthesis of branched polysiloxane grafts on silica surfaces has recently been described [80]. This chemistry involves grafting a living anionic siloxane polymer onto a surface that contains reactive Si – Cl groups (Scheme 7). The living polymers were prepared by anionic ring opening polymerization of cyclotrisiloxane monomer(s) using butyllithium initiator. By taking advantage of reactivity ratios for the anionic copolymerization of monomers **17** and **18**, a copolymer grafting agent was prepared that had functional vinyl groups more concentrated at one end of the graft oligomer. Thus, the resulting grafted surface contained more vinyl groups on the periphery of the graft. The new vinyl groups of these grafted surfaces were then modified to include more Si – Cl groups by a second platinum catalyzed hydrosilylation using HSiCl(CH$_3$)$_2$. Branched polysiloxane grafts on silica particles were prepared by treating this activated surface with yet another batch of living anionic siloxane copolymer grafting agent. This chemistry was been extended to the synthesis of branched polysiloxane grafts with phosphine ligands for im-

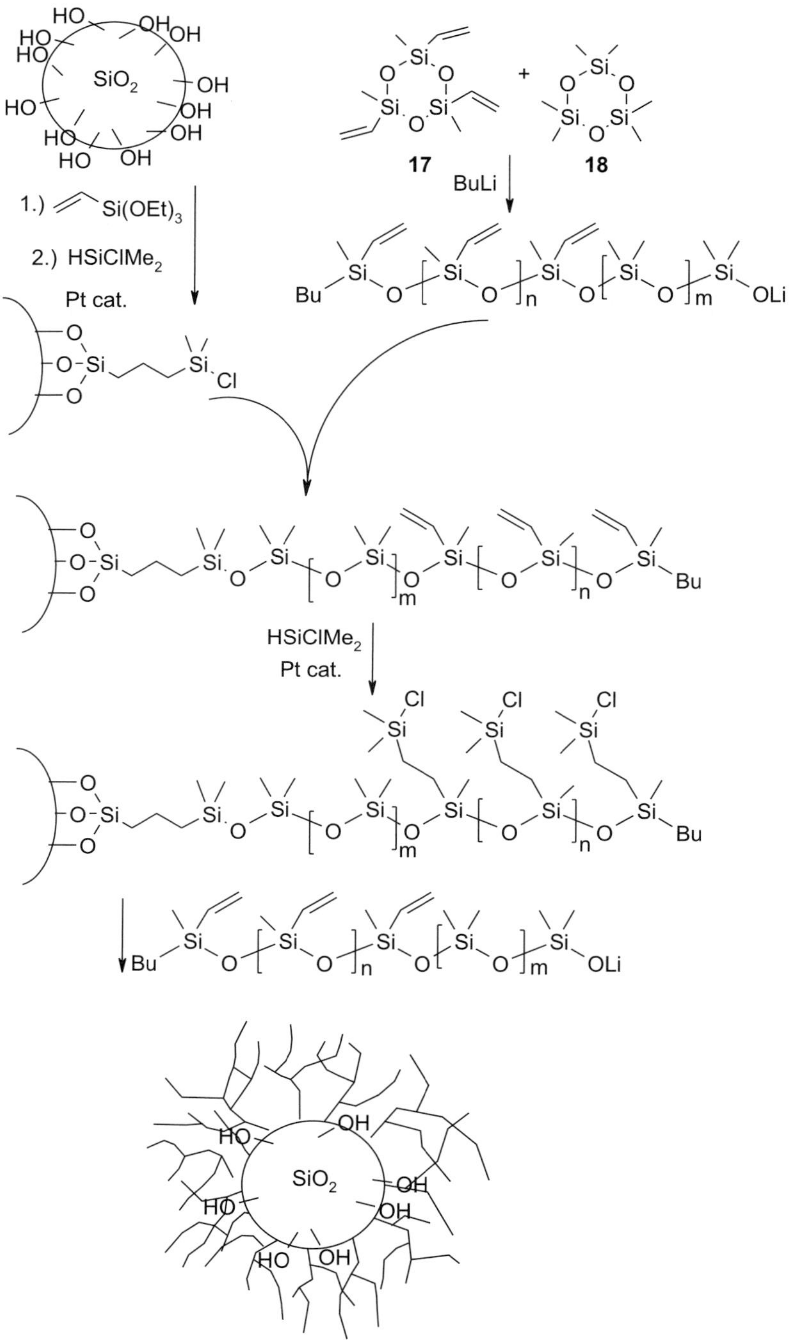

Scheme 7 Synthesis of hyperbranched polysiloxane grafts on silica

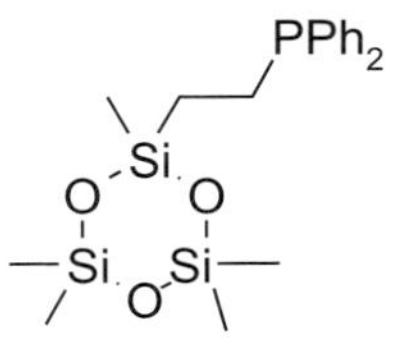

PPh$_2$

19

mobilization of transition metal catalysts [81, 82]. The living anionic siloxane polymer containing phosphine groups necessary for this latter chemistry was synthesized by ring opening polymerization of **19**.

5.2
Dendritic Hyperbranched Grafts of Pd(II) Coordination Polymers

Pincer compounds with a general structure like **20** are common in organometallic chemistry [832]. They are useful as precursors to catalysts and as components of hybrid organometallic polymers [84–86]. In most cases, they have high stability to oxygen and water in addition to high thermal stability.

A recent report described the use of a pincer compound like the so-called SCS-Pd(II) complexe **20** in formation of dendritic grafts on gold surfaces [88]. In this chemistry, the organometallic dendrimer **21** was prepared in solution and then immobilized on gold as part of a self assembled monolayer. The di-alkyl sulfide group on the dendrimer binds to the surface at defects in the self assembled monolayer surface. Tapping-mode atomic force micrscopy was used to show that these organometallic dendrimers were incorporated onto the gold surface with their concentration on gold being dependent on the length time the self assembled monolayer was exposed to **21**.

Similar SCS-Pd(II) complexes have also been used to prepare isolated nanometer-sized objects within a decanethiol self assembled monolayer on gold [88]. In this chemistry, the thioether derivative of a pyridine complex of the SCS-Pd(II) complex **22** was first prepared and immobilized on gold as part of a self assembled monolayer. Then the much better phosphine ligand

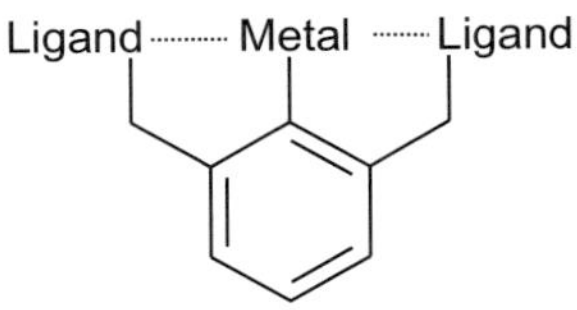

Ligand ········ Metal ·······Ligand

20

23 was added to form a dendritic graft (Eq. 20). This reaction was facilitated by the fact that the necessary ligand exchange chemistry is quantitative and rapid. Extensions of this chemistry would lead to Pd-based organometallic dendrimers on gold. In this initial report, the isolated complexes **24** were characterized by tapping-mode atomic force microscopy. The height of the dendritic graft that was exposed above the decanethiol monolayer was 4.3 nm, in rough agreement with computer modeling that predicted a height for the dendritic unit of 3.4 nm.

(20)

6
Dendrimer Analogs as Hyperbranched Grafts

Hyperbranched poly(amidoamidoamine) (PAMAM) polymers have been grafted on carbon black [89], chitosan [90], glass fibers [91], silica [92–95] and polyethylene [96]. This grafting chemistry introduces highly branched surface immobilized polymers that are better described as hyperbranched polymers rather than dendrimers based on the evidence of incomplete reactions as shown in Scheme 8. The hyperbranched polyamidoamine grafts were synthesized by a two step process from an amine functionalized surface. The hyperbranched grafting in this case involved Michael addition of methyl acrylate to the amine-functionalized surface. Amidation of the resulting surface-bound esters with ethylenediamine then formed a new amine-rich surface. The results were surfaces with amine-rich grafts that could be ana-

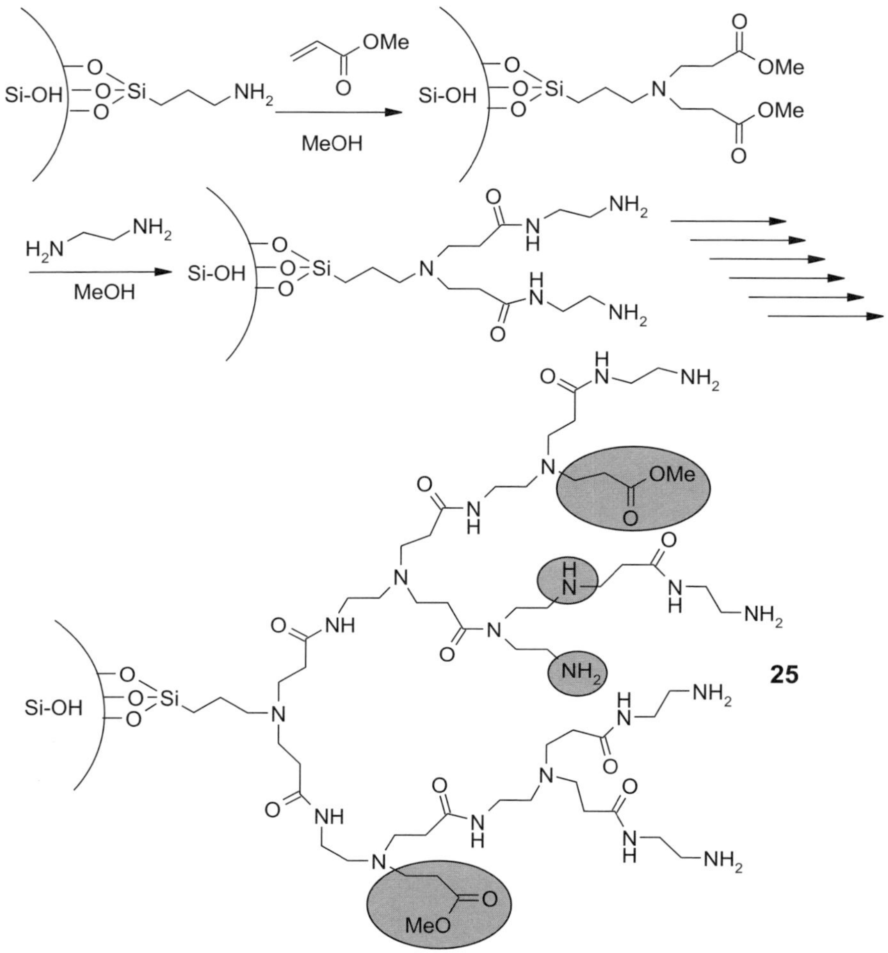

Scheme 8 PAMAM graft synthesis with incomplete branches highlighted in *gray*

lyzed by various methods. For example, beginning with amine-functionalized silica that was assayed by titration to have 0.4 mmol – NH$_2$/g, a tenth generation graft was prepared and analyzed by thermogravimetic analysis (TGA). At this stage, the measured graft loading was ca. 6% of the theoretical loading [92]. Higher loadings relative to the theoretical amount of grafting were seen with lower generation grafts. TGA was only useful for thermally robust silica and glass fiber substrates. Titrimetric analysis of the primary amines was more generally useful and was used to measure graft loadings on silica, glass fiber, carbon black. Titrimetry also was reportedly used for a polyethylene film. For the polysaccharide chitosan and polyethylene films, the graft loadings were also measured by gravimetric analysis.

These amine terminated hyperbranched PAMAM grafts on chitosan [90] and glass fiber [91] have been used to support further grafting chemistry. For example, a living poly(isobutyl vinyl ether) (PIBVE) was prepared using a HCl/ZnCl$_2$ initiator system. This living polymer was then added to a graft like **25** on glass fibers. Presumably the grafting occurs by reaction of the cationic centers in the living cationic polymer with the terminal amines of **25** to form aminals or imines. TGA analysis of the PIBVE grafted on an eighth

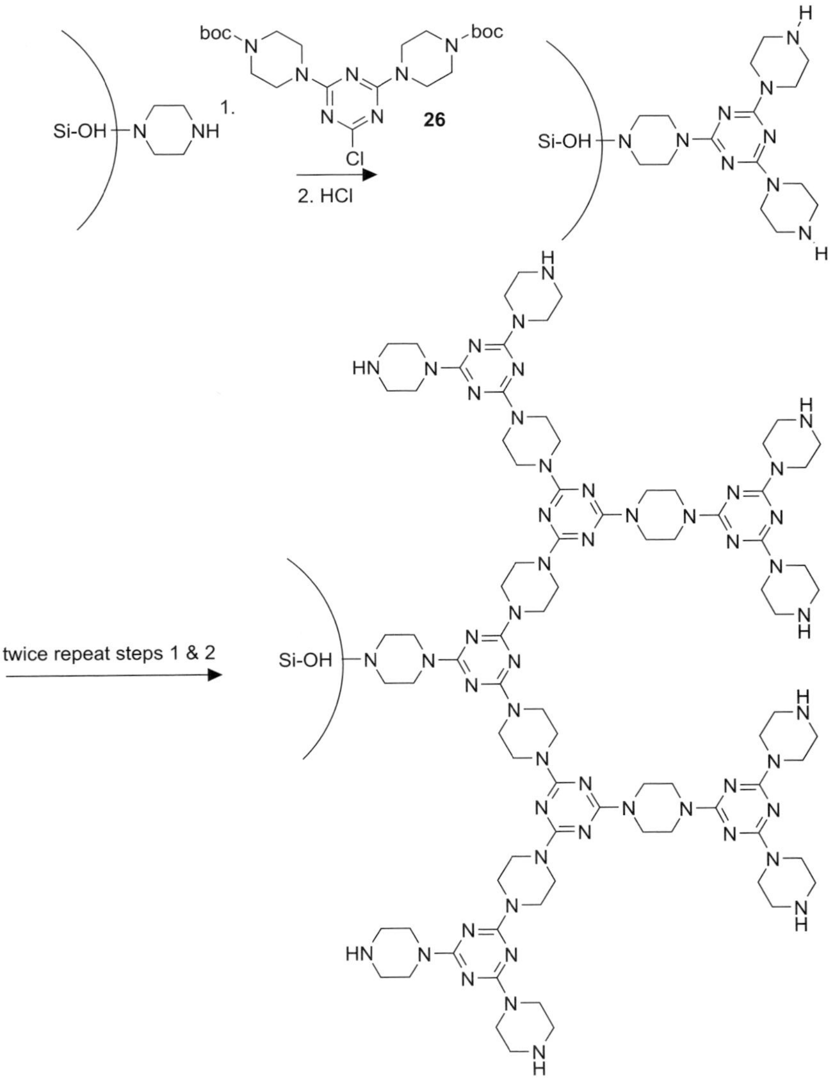

Scheme 9 Synthesis of three generation melamine dendrimer on silica surfaces

generation PAMAM graft **25** on glass fiber showed a 5.2% increase in weight (with an overall grafting of 16.2 weight percent).

Simanek's group has recently compared two methods of grafting melamine-based dendrimers to piperazine functionalized silica [97]. The first method involves the stepwise grafting of a three generation melamine-based dendrimer shown in Scheme 9. This chemistry begins by treating with piperazine functionalized silica with **26**. The *tert*-Boc protecting group was removed by HCl treatment. The grafting and hydrolysis steps were repeated two more times producing a grafted third generation dendrimer. The second grafting approach involved directly grafting a third generation melamine-based dendrimer that was prepared in solution to silica. The progress of either the monomer growth of melamine dendrimers on silica or grafting of a dendrimer to silica was followed by ATR-IR spectroscopy, XPS analysis, matrix-assisted laser desorption/ionization mass spectrometry (MALDI-MS), and TGA analysis. While ATR-IR spectroscopy, XPS analysis, and TGA analysis were all consistent with grafting, the MALDI-MS spectra after HF digestion of the silica supports proved to be the most useful characterization technique. For example, the MALDI-MS spectrum of a third generation melamine-based dendrimer that was synthesized in a stepwise fashion showed peaks that indicated incomplete branching. Peaks were observed that corresponded to the grafted dendrimer and that were consistent with a dendrimer missing one, two, and three branches were detected. In contrast, the MALDI-MS spectrum of grafts prepared using the soluble third generation dendrimer and silica only showed peaks for the third generation dendrimer.

7
Conclusions

Irregularly hyperbranched grafts provide a useful way to modify surfaces. A variety of chemistry can be used and a wide variety of grafts can be prepared. The hyperbranched grafts can serve as supported membranes, as catalyst supports or as substrates for further covalent graft chemistry. Functional groups within these interfaces can be readily modified by solution-state chemistry. The interfaces themselves can be used as media for further chemistry within the interface or as substrates in molecular recognition and self assembly of other macromolecules.

References

1. Mori H, Seng DC, Zhang MF, Müller AHE (2002) Langmuir 18:3682
2. Mori H, Böker A, Krausch G, Müller AHE (2001) Macromolecules 34:6871
3. Nakayama Y, Sudo M, Uchida K, Matsuda T (2002) Langmuir 18:2601

4. Lee HJ, Nakayama Y, Matsuda T (1999) Macromolecules 32:6989
5. Takeuchi Y, Fujiki K, Tsubokawa N (1998) Polym Bull 41:85
6. Tsubokawa N, Hayashi S, Nishimura J (2002) Prog Org Coatings 44:69
7. Hayashi S, Fujiki K, Tsubokawa N (2000) React Funct Polym 46:193
8. Sidorenko A, Zhai XW, Simon F, Pleul D, Tsukruk VV (2002) Macromolecules 35:5131
9. Sidorenko A, Zhai XW, Greco A, Tsukruk VV (2002) Langmuir 18:3408
10. Tully DC, Frechet JMJ (2001) Chem Commun:1229
11. Casado CM, Cuadrado I, Moran M, Alonso B, Garcia B, Gonzalez B, Losada J (1999) Coord Chem Rev 186:53
12. Kriesel JW, Tilley TD (2001) Adv Mater 13:1645
13. Crooks RM, Ricco AJ (1998) Acc Chem Res 31:219
14. Fail CA, Evenson SA, Ward LJ, Schofield WCE, Badyal JPS (2002) Langmuir 18:264
15. Minelli C, Blondiaux N, Losson M, Liley M, Jeney S, Hinderling C, Pugin R, Joester D, Diederich F, Vancso J, Hempenius M, Heinzelmann H (2003) Chimia 57:646
16. Zhou YF, Bruening ML, Bergbreiter DE, Crooks RM, Wells M (1996) J Am Chem Soc 118:3773
17. Bergbreiter DE, Tao GL, Kippenberger AM (2000) Org Lett 2:2853
18. Bergbreiter DE, Xu GF, Zapata C (1994) Macromolecules 27:1597
19. Bergbreiter DE, Jing Z (1992) Journal of Polym Sci Part A, Polym Chemistry 30:2049
20. Kleinfeld ER, Ferguson GS (1996) Chem Mater 8:1575
21. Moad G, Solomon DH (1995) The Chemistry of Free Radical Polymerization. Pergamon, Oxford
22. Bruening ML, Zhou YF, Aguilar G, Agee R, Bergbreiter DE, Crooks RM (1997) Langmuir 13:770
23. Lackowski WM, Franchina JG, Bergbreiter DE, Crooks RM (1999) Adv Mater 11:1368
24. Zhao MQ, Zhou YF, Bruening ML, Bergbreiter DE, Crooks RM (1997) Langmuir 13:1388
25. Zhou YF, Bruening ML, Liu YL, Crooks RM, Bergbreiter DE (1996) Langmuir 12:5519
26. Zhao MQ, Bruening ML, Zhou YF, Bergbreiter DE, Crooks RM (1997) Isr J Chem 37:277
27. Dermody DL, Peez RF, Bergbreiter DE, Crooks RM (1999) Langmuir 15:885
28. Peez RF, Dermody DL, Franchina JG, Jones SJ, Bruening ML, Bergbreiter DE, Crooks RM (1998) Langmuir 14:4232
29. Franchina JG, Lackowski WM, Dermody DL, Crooks RM, Bergbreiter DE, Sirkar K, Russell RJ, Pishko MV (1999) Anal Chem 71:3133
30. Bergbreiter DE, Tao GL, Franchina JG, Sussman L (2001) Macromolecules 34:3018
31. Bergbreiter DE, Tao CL (2000) J Polym Sci Part A, Polym Chem 38:3944
32. Zhang LJ, Shulman MA, Whitesides GM, Grabowski JJ (1990) J Am Chem Soc 112:7069
33. Decher G (1997) Science 277:1232
34. Crooks RM (2001) Chemphyschem 2:644
35. Aoki A, Ghosh P, Crooks RM (1999) Langmuir 15:7418
36. Lackowski WM, Ghosh P, Crooks RM (1999) J Am Chem Soc 121:1419
37. Ghosh P, Crooks RM (1999) J Am Chem Soc 121:8395
38. Ghosh P, Amirpour ML, Lackowski WM, Pishko MV, Crooks RM (1999) Angew Chem Int Ed 38:1592
39. Rowan B, Wheeler MA, Crooks RM (2002) Langmuir 18:9914
40. Amirpour ML, Ghosh P, Lackowski WM, Crooks RM, Pishko MV (2001) Anal Chem 73:1560
41. Ghosh P, Lackowski WM, Crooks RM (2001) Macromolecules 34:1230
42. Bergbreiter DE, Franchina JG, Kabza K (1999) Macromolecules 32:4993

43. Tao GL, Gong AJ, Lu JJ, Sue HJ, Bergbreiter DE (2001) Macromolecules 34:7672
44. Shoichet MS, Mccarthy TJ (1991) Macromolecules 24:982
45. Mammen M, Choi SK, Whitesides GM (1998) Angew Chem Int Ed 37:2755
46. Bergbreiter DE, Srinivas B, Gray HN (1993) Macromolecules 26:3245
47. McDowall DJ, Gupta BS, Stannett VT (1984) Prog Polym Sci 10:1
48. Bergbreiter DE, Liu ML (2001) J Polym Sci, Part A Polym Chem 39:4119
49. Egelhaaf HJ, Olkkrug D, Gebauer W, Sokolowski M, Umbach E, Fisher T, Bäuerle P (1998) Opt Mater 9:59
50. Kippenberger AM (2004) PhD Thesis, Texas A&M University
51. Comina PJ, Beck AK, Seebach D (1998) Org Proc Res Dev 2:18
52. Bergbreiter DE, Chen BS, Lynch TJ (1983) J Org Chem 48:4179
53. Bergbreiter DE, Kippenberger AM, Tao GL (2002) Chem Commun:2158
54. Liu YL, Zhao MQ, Bergbreiter DE, Crooks RM (1997) J Am Chem Soc 119:8720
55. Zhao MQ, Liu YL, Crooks RM, Bergbreiter DE (1999) J Am Chem Soc 121:923
56. Liu YL, Bruening ML, Bergbreiter DE, Crooks RM (1997) Angew Chem Int Ed 36:2114
57. Perez GP, Yelton WG, Cernosek RW, Simonson RJ, Crooks RM (2003) Anal Chem 75:3625
58. Dai JH, Sullivan DM, Bruening ML (2000) Ind Eng Chem Res 39:3528
59. Lee PH, Sawan SP, Modrusan Z, Arnold LJ, Reynolds MA (2002) Bioconjugate Chem 13:97
60. Steffensen MB, Simanek EE (2003) Org Lett 5:2359
61. Jordan R, Ulman A (1998) J Am Chem Soc 120:243
62. Yoon KR, Chi YS, Lee KB, Lee JK, Kim DJ, Koh YJ, Joo SW, Yun WS, Choi IS (2003) J Mater Chem 13:2910
63. Jeong SY, Kim JY, Yang HD, Yoon BN, Choi SH, Kang HK, Yang CW, Lee YH (2003) Adv Mater 15:1172
64. Yoon KR, Koh YJ, Choi IS (2003) Macromol Rapid Commun 24:207
65. Jordan R, West N, Ulman A, Chou YM, Nuyken O (2001) Macromolecules 34:1606
66. Harada Y, Girolami GS, Nuzzo RG (2003) Langmuir 19:5104
67. Juang A, Scherman OA, Grubbs RH, Lewis NS (2001) Langmuir 17:1321
68. Buchmeiser MR, Sinner F, Mupa M, Wurst K (2000) Macromolecules 33:32
69. Weck M, Jackiw JJ, Rossi RR, Weiss PS, Grubbs RH (1999) J Am Chem Soc 121:4088
70. Kim NY, Jeon NL, Choi IS, Takami S, Harada Y, Finnie KR, Girolami GS, Nuzzo RG, Whitesides GM, Laibinis PE (2000) Macromolecules 33:2793
71. Li XM, Huskens J, Reinhoudt DN (2003) Nanotechnology 14:1064
72. Khan M, Huck WTS (2003) Macromolecules 36:5088
73. Kim CO, Cho SJ, Park JW (2003) J Colloid Interface Sci 260:374
74. Kim HJ, Moon JH, Park JW (2000) J Colloid Interface Sci 227:247
75. Moon JH, Kim JH, Kim K, Kang TH, Kim B, Kim CH, Hahn JH, Park JW (1997) Langmuir 13:4305
76. Gomez-Romero P (2001) Adv Mater 13:163
77. Sharp KG (1998) Adv Mater 10:1243
78. Hagrman PJ, Hagrman D, Zubieta J (1999) Angew Chem Int Ed 38:2639
79. Errington J, Mcdonald WS, Shaw BL (1980) J Chem Soc Dalton Transactions: 2312
80. Chojnowski J, Cypryk M, Fortuniak W, Scibiorek M, Rozga-Wijas K (2003) Macromolecules 36:3890
81. Michalska ZM, Rogalski L, Rozga-Wijas K, Chojnowski J, Fortuniak W, Scibiorek M (2004) J Mol Catal A Chem 208:187
82. Rozga-Wijas K, Chojnowski J, Fortuniak W, Scibiorek M, Michalska Z, Rogalski L (2003) J Mater Chem 13:2301

83. van der Boom ME, Milstein D (2003) Chem Rev 103:1759
84. Bergbreiter DE, Osburn PL, Liu YS (1999) J Am Chem Soc 121:9531
85. Pollino JM, Weck M (2002) Org Lett 4:753
86. Albrecht M, van Koten G (2001) Angew Chem Int Ed 40:3750
87. Huisman BH, Schönherr H, Huck WTS, Friggeri A, van Manen HJ, Menozzi E, Vancso GJ, van Veggel FCJM, Reinhoudt DN (1999) Angew Chem Int Ed 38:2248
88. Friggeri A, van Manen HJ, Auletta T, Li XM, Zapotoczny S, Schönherr H, Vancso GJ, Huskens J, van Veggel FCJM, Reinhoudt DN (2001) J Am Chem Soc 123:6388
89. Tsubokawa N, Satoh T, Murota M, Sato S, Shimizu H (2001) Polym Adv Tech 12:596
90. Tsubokawa N, Takayama T (2000) React Funct Polym 43:341
91. Fujiki K, Sakamoto M, Yoshikawa S, Sato T, Tsubokawa N (1999) Composite Interfaces 6:215
92. Tsubokawa N, Ichioka H, Satoh T, Hayashi S, Fujiki K (1998) React Funct Polym 37:75
93. Yoshikawa S, Satoh T, Tsubokawa N (1999) Colloids Surf A, Physicochem Eng Aspects 153:395
94. Okazaki M, Murota M, Kawaguchi Y, Tsubokawa N (2001) J Appl Polym Sci 80:573
95. Murota M, Sato S, Tsubokawa N (2002) Polym Adv Tech 13:144
96. Nakada S, Sawatari C, Tamura K, Yagi T (2001) Colloid Polym Sci 279:754
97. Acosta EG, Gonzalez SO, Simanek EE (2003) Polym Mater Sci Eng 89:793

Adv Polym Sci (2006) 198: 51–124
DOI 10.1007/12_060
© Springer-Verlag Berlin Heidelberg 2005
Published online: 20 December 2005

Surface-Grafted Polymer Gradients:
Formation, Characterization, and Applications

Rajendra R. Bhat[1] · Michael R. Tomlinson[1] · Tao Wu[1,2] · Jan Genzer[1] (✉)

[1]Department of Chemical & Biomolecular Engineering, North Carolina State University,
Raleigh, North Carolina 27695-7905, USA
Jan_Genzer@ncsu.edu

[2]*Present address:*
Polymers Division, National Institute for Standards and Technology,
Gaithersburg, Maryland, USA

Abstract This review presents to-date progress in the formation of surface-tethered polymer assemblies with gradually varying physico-chemical properties. The typical characteristics of the grafted polymers that may change spatially along the specimen include
molecular weight, grafting density on the substrate, and chemical composition. The
concept of surface-anchored polymer assemblies is employed in several projects, including, study of the mushroom-to-brush transition in surface-tethered polymers, monitoring kinetics of controlled/"living" radical polymerization, synthesis of surface-anchored
copolymers with tunable compositions, and analysis of macromolecular conformations
in weakly charged grafted polyelectrolyte and polyampholyte systems. We also discuss the application of grafted polymer gradient systems in studying three-dimensional
dispersions of nanosized guest objects. In the aforementioned examples, the use of
gradient structures both enables methodical exploration of a system's behavior and fa-
cilitates expeditious data measurement and analysis. Furthermore, we outline methods
leading to the formation of orthogonal gradients—structures in which two distinct gradients traverse in orthogonal directions. We illustrate the applicability of molecular
weight/grafting density orthogonal gradients in organizing nanoparticles and controlling protein adsorption on polymer surfaces. Finally, we identify several areas of science
and technology, which will benefit from further advances in the design and formation of
gradient assemblies of surface-bound polymers.

Keywords Gradient · Polymer brush · Self-assembly · Nanoparticles · Protein adsorption

Abbreviations

3-D	three-dimensional
ATRP	atom transfer radical polymerization
BMPUS	[11-(2-bromo-2-methyl)propionyloxy] undecyltrichlorosilane
CMPE	1-trichlorosilyl-2-(m/p-chloromethylphenyl) ethane
DI	deionized
[DS]	degree of swelling
EGF	epidermal growth factor
FTIR	Fourier transform infrared spectroscopy
GPS	3-glycidoxypropyl trimethoxysilane
h	dry thickness of surface-anchored polymer
H	wet thickness of surface-anchored polymer
$[H^+]$	proton concentration in bulk solution
H_b	height of a grafted polymer in a brush conformation
H_{DIW}	polymer wet thicknesses in "pure" water
$H(IS)$	polymer wet thicknesses evaluated at a given IS
H_m	height of a grafted polymer in a mushroom conformation

IEP	isoelectric point
IS	solution ionic strength
IS_{max}	solution ionic strength at the transition from OB to SB
K_a	equilibrium constant of the acid-base equilibrium
MMA	methyl methacrylate
M_n	number average molecular weight of polymer
MW	molecular weight
N	degree of polymerization
N_A	Avogadro's number
NB	neutral brush
NEXAFS	near-edge X-ray absorption fine structure
OB	osmotic brush
OTS	n-octyltrichlorosilane
P2VP	poly(2-vinyl pyridine)
PAA	poly(acrylic acid)
PAAm	poly(acryl amide)
PDMAEMA	poly(dimethyl aminoethyl methacrylate)
PE	polyethylene
PEG	poly(ethylene glycol)
PEG-MA	poly(ethylene glycol) methacrylate
PEO	poly(ethylene oxide)
PEY	partial electron yield
PGMA	poly(glycidyl methacrylate)
PHEMA	poly(2-hydroxylethyl methacrylate)
PMAA	poly(methacrylic acid)
PMMA	poly(methyl methacrylate)
PO	paraffin oil
PS	polystyrene
PtBA	poly(*tert*-butyl acrylate)
R_g	radius of gyration of a polymer
R_p	rate of polymerization
SAM	self-assembled monolayer
SB	salted brush
SE	spectroscopic ellipsometry
SFM	scanning force microscopy
tBA	*tert*-butyl acrylate
T_g	glass transition temperature of a polymer
XPS	X-ray photoelectron spectroscopy
α	degree of dissociation in bulk solution
α_0	"internal" degree of dissociation
Δ, Ψ	ellipsometric angles related to the change of amplitude and phase shift
θ_{DIW}	contact angle of deionized water
λ	wavelength of light
v	excluded volume parameter
v_e	electrostatic excluded volume parameter
ρ	polymer density
$\rho_{s,\infty}$	salt concentration in bulk solution
σ	polymer grafting density
φ	polymer volume fraction
χ	Flory–Huggins interaction parameter

1
Introduction

Polymer coatings offer an efficient and convenient way of modifying physico-chemical characteristics of material surfaces [1, 2]. Such polymer-modified surfaces are used extensively in a variety of applications, including photolithographic masks [3], adhesion promoters [4], lubricants [5], stabilizers of colloidal particles [6], nonfouling coatings [7–9], responsive materials [10], and others. Polymers serve as excellent candidates for surface modification on account of several reasons: (1) compared to low molecular weight coatings such as self-assembled monolayers (SAM) polymers have better and more tailorable mechanical properties; (2) they can form thick films thus providing a large number of functional groups; (3) they offer a wide variety of functional groups to choose from and; (4) they can serve as multifunctional stimuli-responsive materials. Traditionally, polymer coatings are created by physically attaching a polymer layer to the surface of interest. Typical methods of forming physisorbed polymer coatings include spin casting or solution dip-coating. The nature of interaction between the polymer and the surface in these coatings is usually noncovalent. Although physisorbed polymer coatings serve their purpose immediately after the fabrication process, their performance diminishes over a protracted time period due to the erosion of the coating under harsh application conditions. To overcome these shortcomings, it is deemed necessary to chemically bind the polymers to the surface.

The chemisorbed polymer structures are, in turn, further classified as either "grafted onto" or "grafted from" polymers, the main difference between the two being the procedure by which macromolecules are attached to the surface (cf. Fig. 1). In grafted-onto coatings, preformed polymer chains are simply attached to a given surface via chemical reaction between the functional groups present on the surface and those along polymer chains (usually end-functionality). The advantage of such polymer assembly is that the molecular weight and consequently the chain length of grafted polymer are well-characterized. However, with this approach it is difficult to form thick layers (thicker than a few tens of nanometers) and the coating suffers from a low density of grafted chains due to the steric hindrance for the attachment of new polymer chains once a few initial chains have been grafted. To overcome these crowding effects and increase the density of attached chains, monomers are polymerized from surface-bound polymerization initiators using the grafting-from approach. Since the diffusing species in the grafting-from approach are monomer molecules, a large number of chains can be polymerized from the surface-attached initiators, thus achieving a high surface density of the chains. Advantages of the grafting-from method over the grafting-onto technique include: (1) very thick and dense coatings can be routinely created; and (2) a wide variety of monomers can be polymer-

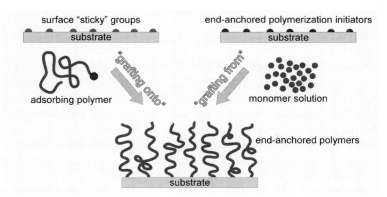

Fig. 1 Schematic illustrating the formation of surface-anchored polymer assemblies by utilizing the "grafting onto" and "grafting from" methods

ized since the monomer or polymer need not have any specific functional groups as is required in the grafting-onto method. However, the molecular weight and the chain length distributions of polymer chains formed by the grafting-from methodology cannot always be accurately controlled and measured. Additional complications arise from the fact that the substrate geometry also influences the rate of polymerization and the chain length distribution. For example, while the kinetics of polymerization for the grafting-from reactions originating from the surfaces of nanoparticles does not seem to be too different from the typical polymerization in the bulk [11, 12], switching to flat or even concave geometries has a pronounced effect on the rate of polymerization and chain length distribution (Tomlinson et al., 2004, unpublished results). Finally, when growing multifunctional macromolecules, such as copolymers, the activity of the growing chain, which participates in further polymerization, the so-called macroinitiator, may depend on its length and chemistry. In spite of these limitations, grafting-from methodologies have recently attracted considerable attention within the polymer community [13].

Conformations of tethered polymer chains are significantly different from those of free polymers in solution [14–16]. The main parameters governing the conformation of macromolecules in solution include the quality of the solvent (expressed in terms of the monomer–solvent interaction parameter), chain stiffness, and the degree of polymerization, N. In good solvents, where favorable interaction between monomer and solvent molecules dominate the loss of entropy due to chain stretching, polymer molecules assume a relaxed conformation in the form of an expanded coil. In contrast, a poor solvent causes monomers to minimize contacts with solvent molecules as much as possible, consequently resulting in the formation of a collapsed globule. Radius of gyration of a polymer chain, which is a measure of the average size of the chain, is given by $R_g \sim N^{3/5}$ under good solvent conditions. This scaling relationship shows a much weaker dependence on N (or equivalently molecular weight) under poor solvent conditions with $R_g \sim N^{1/3}$. In the so-called theta solvents, which are intermediate between the good and poor solvents, the chain dimension scales as the unperturbed radius of gyration, $R_g \sim N^{1/2}$. Grafting of polymer chains to a surface dramatically modifies the way in which the constraints of monomer–monomer interaction and chain stretching on polymer conformation are manifested. In addition to N, the conformation of end-tethered polymer chains is governed by the number of polymer chains grafted per unit area of the substrate characterized by grafting density of chains on the surface, σ. When a few long polymer chains are attached to the substrate, they do not overlap and as long as there is no special interaction with the substrate, the conformation of these chains is similar to that in solution. Under good solvent conditions, the chains try to maximize the number of contacts with the solvent molecules while keeping chain stretching to a minimum. This behavior results in the formation of a so-called mushroom conformation (cf. Fig. 2). Wet thickness of a polymer in the mushroom regime under good solvent conditions has been theoretically predicted and experimentally verified to scale as $H \sim N\sigma^0$ [14, 15]. It must be pointed out that the aforementioned scaling relations apply only to neutral grafted polymers, i.e., the polymers are devoid of any chargeable functional groups along their backbone. As the grafting density of chains on the surface increases, the osmotic pressure among the chains increases and this forces the chains to stretch normal to the substrate. This equilibrium conformation, in which polymer chains are stretched away from the grafting surface, has been coined as a polymer brush (cf. Fig. 2) [15]. The wet thickness of the brush in a good solvent scales as $H \sim N\sigma^{1/3}$. Thus, compared to free polymer chains in solution, the size of grafted chains exhibits a stronger dependence on N and an additional dependence on grafting density. When placed in poor solvent conditions, the surface-tethered polymers collapse, giving rise to collapsed chain conformations. While at small σ the individual chains remain isolated

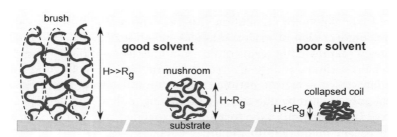

Fig. 2 Schematic representation of surface-anchored polymers in brush (*left*) and mushroom (*middle*) conformations in good solvents. Also shown is the conformation of a surface-tethered polymer under poor solvent conditions (*right*)

(cf. Fig. 2), with increasing σ they form surface-bound collapsed aggregates of various shapes and sizes [14, 17].

As the conformation of the grafted polymer chains governs the effectiveness of the polymer coating for a given application, it is desirable to be able to manipulate the conformation by controlling N and σ of the grafted chains. For example, it has been predicted that adsorption of small proteins (and the subsequent bio-fouling) on a surface can be minimized by using polymer brushes having high grafting density whereas thicker brushes (high N) are preferred to achieve similar repellency for larger proteins [8]. Polymer brushes are widely used to prevent coagulation of colloidal particles. This steric stabilization is best realized by using long grafted chains with a relatively high σ [6]. Adhesion promotion between a polymer melt and brush-coated surface is found to depend critically on σ [18]. Short chains with very high σ are suitable for enhancing the lubrication of a polymer-coated surface with a sliding surface [5]. In order to gain a complete understanding of any of the above-mentioned phenomena, it is imperative that properties of the polymer surface as well as those of the environment in contact with the polymer surface be systematically varied and their effect on the behavior of the system be studied. There are two ways by which properties of a polymer surface can be studied: (a) prepare a large number of polymer samples with each sample having different surface properties; or (b) design and fabricate a single sample in which properties of the polymer surface vary continuously along the sample between two extreme values. Apart from being more time-consuming and expensive, the former approach involves changing individual variables in a discrete and uncontrollable manner, which does not guarantee exploration of the entire behavioral spectrum of the system. These deficiencies can be minimized or even completely eliminated in some cases by utilizing the latter approach, wherein polymer gradient surfaces are generated and utilized.

Gradient surfaces offer powerful avenues enabling systematic variation of one or more brush properties. Employing a gradient surface to study a complex, multivariate phenomenon enables unambiguous interpretation of the system's response to a given stimulus. Since all other properties of the system remain constant, the system's response can be unequivocally attributed to the gradually changing surface property along the gradient. Additionally, gradient surfaces offer combinatorial platforms for quick and inexpensive investigation of the multivariate phenomenon [19, 20]. Similar study by traditional methods typically requires preparation of numerous samples, ostensibly under similar experimental conditions. Various types of surface-grafted polymer gradients are illustrated in Fig. 3 (homopolymer gradients) and Fig. 4 (copolymer, mixed polymer gradients). These can be gradients in polymer chain length, grafting density, composition (lateral as well as transverse) or a combination of any of these parameters. These polymer gradients not only work as combinatorial substrates for investigation of a complex phenomenon but they also serve as soft-matter templates, thus allowing fabri-

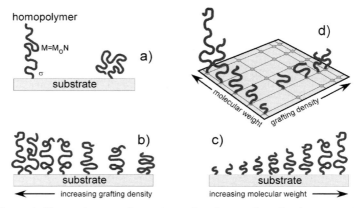

Fig. 3 Schematic illustrating **a** conformations of surface-anchored polymers, and polymer brush assemblies with a **b** grafting density gradient, and **c** a gradient in polymer length. Part **d** depicts polymer conformations on a substrate comprising grafting density and polymer length orthogonal gradients

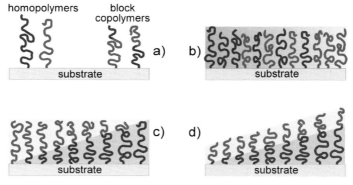

Fig. 4 Schematic illustrating **a** conformations of grafted homopolymer brushes made of two chemically distinct polymers and diblock copolymers, **b** mixed homopolymer brushes with gradually changing grafting densities of two polymers, **c** grafted diblock copolymers with a constant length and gradually changing composition, and **d** grafted diblock copolymers with a gradually changing length of one block and a constant length of the other block

cation of gradient assemblies of nonpolymeric objects. From a materials point of view, polymer gradients offer a unique structure wherein physico-chemical characteristics such as wetting, or chemical composition change gradually, typically over a broad range of properties. Such surfaces with unidirectional variation of material properties may be useful in directing motion of liquid droplets or nano-objects thus acting as molecular motors. Some of these applications are addressed in greater detail towards the end of this review.

While the field of surface-bound polymer gradients is still in its nascent stages, it has a great potential for further exploration. Accordingly, the goal of this review is to familiarize the reader with existing methodologies that enable the formation and characterization of surface-attached polymer gradients, with an emphasis on their utility to study complex phenomena. Specifically, we will elaborate on the role of gradient structures in advancing basic knowledge of polymer behavior in confined spaces as well as fabricating functional devices. We restrict the review to polymer gradients that are chemically bound to a substrate. Although industrially relevant, we will not venture into the area of gradient polymer coatings that lack a chemical bond between polymer chains and substrate [19]. This review is organized as follows: In Sect. 2, we describe the existing techniques of forming end-tethered polymer gradients; in Sects. 3 and 4 we review work relevant to polymer surfaces with a gradient in grafting density of attached chains and their conformational response to ionic solutions; Sects. 5 and 6 deal with surfaces having a gradient in polymer chain length and composition, respectively. In Sect. 7, we report on the solution behavior of grafted polyampholytes prepared by using the techniques described in Sects. 5 and 6. Section 8 illustrates how one can combine two different kinds of gradients to generate orthogonal gradient structure. Finally, Sect. 9 covers applications of the grafted polymer gradients in two rapidly growing disciplines of chemical sciences, viz. nanotechnology and biomaterials. The Outlook section concludes this review by pinpointing several areas of science and technology, which we believe will greatly benefit from the application of gradient polymer substrates.

2
Review of Techniques of Preparation and Characterization of Polymer Gradients

Although a multitude of methodologies exist for the creation of gradient assemblies of short organic modifiers [21], relatively few techniques are available for generating gradients of surface-bound polymers. Most of them rely on selective physical or chemical "cure" of surfaces before or during growth/attachment of polymer. These treatments include creation of density gradient of surface sites used for adsorbing or growing polymers, gradual immersion or withdrawal of a substrate from a polymerization solution, regulation of radiation intensity during photoimmobilization, exposing a substrate to a temperature gradient etc. A brief review of these strategies is presented below.

2.1
Surface-Grafted Polymers Through Corona Treatment

In this technique, radio-frequency corona discharge treatment was first used to create oxygen-rich moieties on a polyethylene (PE) substrate [22, 23]. The density of these moieties on a PE sheet was varied by continuously increasing the power of the corona discharge from a knife-type electrode as the substrate was gradually exposed to the electrode. The PE substrate was then dipped into a monomer solution maintained at an elevated temperature. At this temperature, oxygen-rich moieties on the surface decomposed into free radicals, which served as initiators for grafting-from polymerization. The density gradient of oxygen-rich moieties on the substrate thus resulted in a density gradient of grown polymers. Typical monomers used were acrylic acid, sodium p-styrene sulfonate, and N,N-dimethylaminopropyl acrylamide. These gradients were employed to investigate the effect of wettability, functional group density and charge on protein adsorption and cell growth on these surfaces [22, 23]. Although this technique was successful in creating density gradients of functional groups, no information about the molecular parameters of these surface-bound gradients, such as N or σ, was provided.

2.2
Gradients Prepared by the Grafting-Onto Technique

As mentioned previously, the grafting-onto method involves the preparation of substrate-bound polymers wherein pre-formed polymer chains are simply attached to the substrate through a chemical reaction between a functional group on the substrate and a functional group (typically end-group) on the polymer. Lee and coworkers prepared grafting density gradients of comb-like poly(ethylene glycol) (PEG) on a PE substrate (cf. Fig. 5) [24, 25]. To achieve this, the researchers first subjected a PE sheet to a corona discharge treatment, with the corona power increasing gradually along the length of the sheet. As mentioned in Sect. 2.1, this type of corona treatment produces a density gradient of oxygen-rich moieties on the PE surface. At elevated temperatures, these moieties served as binding sites for the attachment of poly(ethylene glycol) methacrylate (PEG-MA) chains. Immersing a PE sheet with a continuous gradient in binding sites in a solution of PEG-MA main-

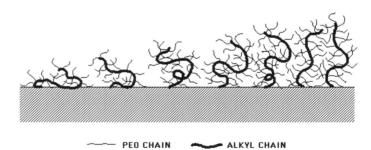

                        ———— PEO CHAIN   ～～～ ALKYL CHAIN

Fig. 5 Schematic diagram showing a comb-like poly(ethylene oxide) gradient produced on a polymer surface. (Reproduced with permission from [24])

tained at high temperature resulted in the formation of grafted polymers with a gradient in the grafting density of the chains. In the PEG-MA gradient, the chain length of PEG-MA remained presumably constant along the substrate; only the density of the chains varied. Since PEG is well known for its protein-resistant nature, Lee et al. used the PEG-comb gradients to study plasma protein and platelet adsorption. They found that as the surface density of PEG chains increased, protein and cell adsorption decreased [25] and the surfaces became more nonfouling. Major advantages of their gradient approach were that it reduced the number of experiments as well as the methodological error associated with those experiments.

Ionov et al. employed a temperature gradient to create a grafting density gradient of surface-attached polymer chains (cf. Fig. 6, left part) [26]. For this purpose, they first either dip-coated a silica wafer with a thin layer of poly(glycidyl methacrylate) or attached a monolayer of epoxy silane, which served as an anchoring layer for further attachment of polymer chains. A layer of end-functionalized polymer was spin-coated on top of the anchoring layer and the whole assembly was heated above the glass transition temperature of spin-cast polymer. The substrate was kept on a heating stage with a temperature gradient along the length of the stage. Temperature was varied from a few degrees below the glass transition temperature (T_g) to tens of degrees above T_g. Because of the temperature-dependent grafting kinetics, the extent of grafting of polymer chains at a given point on the surface depended on the temperature prevailing at that point. Consequently, the temperature gradient along the substrate was translated into a grafting density gradient of anchored chains. The right portion of Fig. 6 shows that the amount of polymer attached increases as grafting temperature increases along the substrate.

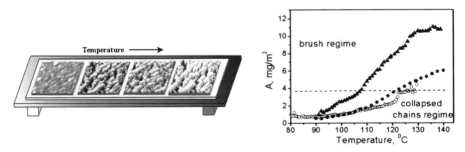

Fig. 6 *(left)* Preparation of ultrathin tethered polymer layers with gradually changing thickness by utilizing the grafting-onto approach using a temperature gradient created on a heating stage. *(right)* Grafted amount of carboxylic acid-terminated polystyrene (PS-COOH) film grafted through *(open circles)* 3-glycidoxypropyl trimethoxysilane (GPS, annealing time = 4 h), *(closed circles)* poly(glycidyl methacrylate) (PGMA, annealing time = 4 h), and *(closed triangles)* PGMA (annealing time = 12 h) anchoring layers. (Reproduced with permission from [26])

The gradient also displayed a continuous change in morphology (cf. Fig. 6, left part) and wetting properties.

Recently, Ionov and coworkers used their gradient formation technique to form a mixed brush grafting density gradient [27]. In this assembly, gradients of two different polymers run counter to each other, i.e., the grafting density of one polymer increases in one direction along the substrate whereas that of the other decreases, and vice versa (cf. Fig. 4b). To create such a two-component polymer gradient assembly, the authors first formed a gradient of poly(*tert*-butyl acrylate) (PtBA) according to the technique described in the above paragraph. This gradient was then "backfilled" by poly(2-vinyl pyridine) (P2VP) by utilizing the same procedure as was used for grafting the first polymer. The resultant structure comprised a mixed polymer brush gra-

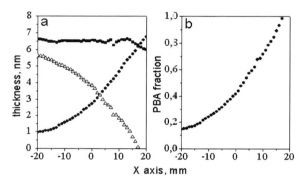

Fig. 7 Ellipsometric mapping of PBA/P2VP gradient brush, total thickness (■), PtBA layer (●), P2VP layer (△). (b) Fraction of PtBA *versus* the point coordinate on the sample (Reproduced with permission from [27])

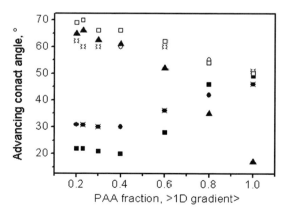

Fig. 8 Contact angle at different pH: 2.2 (■), 2.54 (●), 3.24 (○), 4.95 (□), 9.95 (▲) of PAA/P2VP brush *vs.* composition (Reproduced with permission from [27])

dient with a constant thickness and a variable concentration of the PtBA and P2VP grafted polymers (cf. Fig. 7). After converting PtBA into poly(acrylic acid) (PAA), the wettabilities along the substrate were shown to switch from hydrophilic to hydrophobic upon varying pH (cf. Fig. 8).

2.3
Gradients Prepared by the Grafting-From Technique

The grafting-from method generates surface-bound polymers with a high grafting density due to the minimal diffusion resistance offered to monomer molecules as the polymerization proceeds on the surface. For a given polymer molecular weight (M_n), a higher grafting density of polymers grafted to a surface increases the thickness (h) of the resulting film, $h = \sigma M_n/(\rho N_A)$, where M_n and ρ are molecular weight and density of the polymer, respectively, and N_A is Avogadro's number. Such thick and dense coatings are useful in a number of applications, such as etch-barrier material in photolithography. Although a large variety of polymerization reactions can be used to grow polymers from surface-bound initiators, controlled/"living" polymerizations are preferred because: (1) they retain the simplicity of the radical reactions while producing nearly monodisperse polymers due to fast initiation and minimal termination; (2) high molecular weight polymers can be routinely synthesized due to the controlled nature of the polymerization and; (3) they potentially allow for tailoring the chain microstructure (linear, grafted, star, and others) and facilitate the synthesis of block copolymers [13]. The controlled polymerization techniques that have been applied to create surface-grafted polymer assemblies include [28]: living anionic polymerization [29, 30], living cationic polymerization [31], living ring opening polymerization [32–34], ring opening metathesis polymerization [35, 36], living radical polymerization, mainly nitroxide-mediated polymerization [37], reversible addition-fragmentation chain transfer polymerization [38], and atom transfer radical polymerization (ATRP) [39, 40]. Of all these techniques, ATRP has emerged as a robust and versatile polymerization technique capable of synthesizing a large range of functional polymers on a variety of surfaces [41–44]. With the discovery of ATRP as a controlled/"living" radical polymerization technique, the interest in the field of surface-initiated polymerization has shown a dramatic surge. This is predominantly due to the ease with which ATRP can be carried out on surfaces. Unlike ionic polymerizations, ATRP is not as demanding with regard to monomer and solvent purity, the presence of moisture and oxygen, etc. A large number of functional monomers that do not undergo ionic polymerization, such as water-soluble monomers, can be easily polymerized by ATRP [41, 42, 45, 46]. ATRP is especially powerful in synthesizing polymers with novel topologies, composition profiles, and functionalities on flat as well as curved semiconductor, metal, and polymer surfaces (cf. Fig. 9).

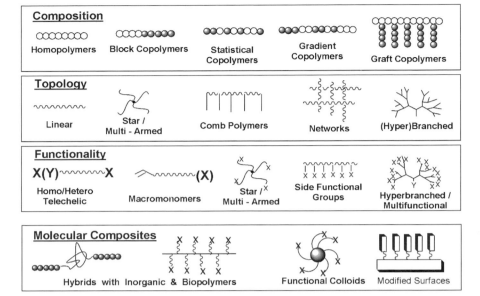

Fig. 9 Schematic representation of controlled topologies, compositions, and functionalities of polymers and molecular composites prepared by atom transfer radical polymerization (ATRP). (Reproduced with permission from [42])

Our group has used ATRP extensively to synthesize a variety of grafted-from polymer gradients. These include gradients in (1) grafting density of polymer chains; (2) polymer molecular weight; (3) polymer chain composition and; (4) a combination of these gradients. While these gradients are schematically illustrated in Figs. 3 and 4, their synthesis, characterization, and applications are described in great detail in subsequent sections.

2.4
Miscellaneous Methods for Immobilizing Polymers in Gradient Patterns

Recently, a few novel approaches involving the formation of gradient grafted polymer assemblies have emerged. For example, Ito and coworkers used a photomask with a gradient pattern to selectively immobilize a biomolecule—epidermal growth factor (EGF)—on a polystyrene tissue culture plate [47]. Specifically, they coated a polystyrene substrate with polyallylamine-tagged EGF and exposed the coating to ultraviolet rays through the gradient photomask. Grafting of EGF took place only on those parts of the substrate that were exposed to UV rays. The EFG gradient pattern was visualized by immunostaining with anti-EGF antibody. The EGF gradient facilitated quantitative evaluation of cell growth as a function of the concentration of immobilized biomolecule. In a similar study, a thermoresponsive polymer was attached to a polystyrene substrate using the gradient micropattern [48]. This gradient was employed to examine the effect of concentration of immobilized thermoresponsive polymer and temperature on the adhesion of fibroblast cells. Hu and coworkers used laser activation to generate a microscale gradient of a protein tagged to a photoactive cross-linking agent [49]. For this purpose, they rastered the laser beam over a protein-coated surface at progressively faster speeds, thereby varying the laser exposure time across the length of the sample. Since the immobilization of protein on the surface was proportional to the laser exposure time, the raster pattern of the beam produced a gradient pattern of immobilized protein.

2.5
Gradient-Graft Copolymers

Graft copolymers are systems, in which one type of polymer (say A) is grafted chemically to the backbone of another polymer (say B). The grafting density of A is dictated by the distribution of the grafting sites along B; these can be perfectly ordered or "perfectly random". Alternatively, there may be a gradient in the density of the grafting sites along B. Growth of polymer A from the backbone having a gradient in grafting sites results in the formation of gradient graft copolymers. Synthesis of such copolymers with gradients in the grafting density of A has recently been reported by the Matyjaszewski group.

Copolymers with a composition gradient are needed as precursors for the synthesis of gradient-graft copolymers. In the composition gradient copolymer the local concentration of one monomer unit along the polymer backbone varies continuously from one end of the chain to the other (cf. Fig. 9) [50–52]. As far as the sequence of the two co-monomers along the backbone is concerned, the gradient composition copolymers can be considered to be intermediate between random and block copolymers. This sequence distribution along the backbone can result either from the differences in the reactivity ratios of the two monomers (spontaneous gradient) or from different rates of addition of the two monomers in the reaction vessel (forced gradient). The reactivity ratio is defined as the ratio of the rate constant of homo-propagation (i.e., tendency of active species to react with species of its own kind) to that of cross-propagation (i.e., predisposition of a monomer species to react with species of another kind in preference to species of its kind). When one monomer tends to homo-propagate but the other prefers to cross-propagate, the resulting copolymer is a gradient composition copolymer, where the two ends of the polymer are enriched with two monomers and the middle part of the polymer has a composition gradient. The compositional breadth of such a spontaneous gradient is determined by the reactivity ratios of the two monomers. To create a forced composition gradient copolymer, the feed rates of one or both monomers are varied either continuously or periodically throughout the course of the reaction. The width of the gradient is predominantly determined by the differences in the feed rates of the two monomers. Properties of the gradient copolymers are significantly different from those of random or block copolymers [50]. Thus, a judicious choice of the two co-monomers and their feed rates can result in the formation of a gradient copolymer having much improved properties compared to the two parent polymers. These gradient copolymers are envisaged to be useful as polymer blend compatibilizers, pressure sensitive adhesives and as novel materials for vibration and noise dampening [53, 54].

As mentioned above, copolymers with a compositional gradient can serve as precursors for generating gradient-graft copolymers, where the number of

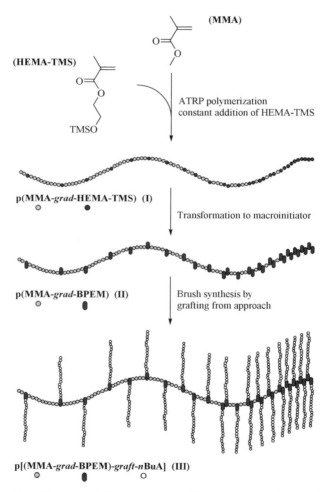

Fig. 10 Synthesis of the macroinitiator precursor (I), the macroinitiator (II) and macromolecular gradient brush copolymer (III). (Reproduced with permission from [55])

side chains attached to the polymer backbone gradually decreases from one end of the backbone to the other [55]. Börner et al. synthesized such gradients according to the scheme shown in Fig. 10. First, they created a forced compositional gradient along the backbone by continuously adding one monomer in the reaction vessel while the other monomer is steadily consumed. One of the two monomers along the backbone then served as an initiator for growing side chains made of a third monomer. Since the backbone itself had a concentration gradient, the grafted copolymer showed a continuous change in side chain grafting density. Such grafted copolymer gradient brushes are highly anisotropic with a bulky head and a thin tail. For further details, the reader is referred to excellent review articles by Matyjaszewski and coworkers dealing with the synthesis of gradient copolymers by controlled radical polymerization techniques [53, 54].

2.6
Methods of Characterizing Gradient Properties

Characterization of gradient substrates involves determining the same physico-chemical characteristics as those measured on "conventional" non-gradient specimens. Since the physico-chemical characteristics of gradient samples vary laterally, a judicious choice of technique is required. Such a choice is governed by the interplay among: (1) the lateral resolution of a given technique; (2) the lateral scanning ability of the technique; and (3) the distance on the surface over which the gradient is present (gradient steepness). Only techniques, whose lateral resolution is better than the gradient steepness can be utilized. Depending on the lateral scanning ability of a given technique vis-à-vis gradient steepness, one can distinguish between two modes of operation. If the lateral scanning ability of the technique is comparable to or slightly larger than the gradient steepness, information content about the gradient substrate can be accessed in a single measurement. We refer to such a mode as "global". If, however, the probe's lateral scanning ability is much smaller than the distance over which the gradient exists, one has to repeat the measurement on various portions of the substrate along the gradient in order to fully characterize the specimen. We call this a "local" mode. In order to illustrate the difference between the two modes let us consider using chemical force microscopy (CFM) [56]. CFM measures the forces between the surface and a chemical moiety, such as end-functionalized alkanethiol, that has been covalently attached to the probing tip. Considering that the largest scanning area accessible with a single CFM measurement is $\approx 125 \times 125\ \mu m^2$, CFM can be used as a global technique on gradient samples whose gradient steepness is equal to or smaller than $\approx 125 \times 125\ \mu m^2$. If the gradient variation occurs over a much larger lateral distance on the substrate, multiple individual measurements with CFM have to be carried out at various positions on the substrate in order to characterize the entire specimen. In the latter case, CFM operates in the local mode. The following paragraphs offer a brief overview of several local techniques that have been utilized to characterize polymer gradients. Some of the most important properties of polymer gradients that need to be characterized are wettability, chemical composition, thickness of the surface layer, and morphology.

Static or dynamic contact angle measurements are the easiest and most widely available methods of determining wettability [57]. In static contact angle experiments, one measures the wetting angle of a small volume of probing liquid on the surface. The typical area on the surface probed by contact angle techniques ranges from hundreds of square micrometers to square millimeters. Performing static contact angle experiments under the condition where the needle and the probing liquid are not separated allows for determining the so-called advancing and receding contact angles. Moreover, keeping the needle in contact with the probing liquid while performing contact angle measurements on wettability gradients ensures that the droplet does not move towards the more wettable region on the substrate. Separating the needle from the liquid would allow the liquid to move on the surface (see [57]), particularly in gradients with steep boundaries between the hydrophobic and hydrophilic regions. The dynamic contact angle (DCA) measurements are usually performed using the Wilhelmy plate. Examples of the DCA measurements on gradient substrates can be found elsewhere [58]. While useful in providing macroscopic level information about the chemistry on the gradient surfaces, contact angle methods are not capable of delivering information about the structural properties of the gradients on a molecular level (such as concentration of a particular chemical group, orientation of molecules, etc).

For acquiring molecular-level information, one has to turn to more sophisticated probes. Ruardy and coworkers discuss the utilization of X-ray photoelectron spectroscopy (XPS) [59] and infrared spectroscopy (IR) for determining concentration of a particular species along molecular gradients [58]. Recently, near-edge X-ray absorption fine structure (NEXAFS) spectroscopy [60] has emerged as a powerful tool for characterizing gradients [61, 62], because it allows for simultaneous investigation of both the surface chemistry and the molecular orientation. NEXAFS involves the resonant soft X-ray excitation of a K or L shell electron to an unoccupied low-lying antibonding molecular orbital of σ or π symmetry, σ^* and π^*, respectively. The initial state K or L shell excitation gives NEXAFS its element specificity, while the final-state unoccupied molecular orbitals provide NEXAFS with its bonding or chemical selectivity. A measurement of the intensity of NEXAFS spectral features enables the identification of chemical bonds and determination of their relative population density within the sample. Because of the fixed geometry between the sample and the X-ray beam and the fact that the $1s \rightarrow \sigma^*$ and $1s \rightarrow \pi^*$ excitations are governed by dipole selection rules, the resonance intensities vary as a function of the direction of the electric vector

E of the incident polarized X-ray relative to the axis of the σ^* and π^* orbitals. This, coupled with the fact that sharp core level excitations for elements C, N, O, and F occur in the soft X-ray spectral region, makes NEXAFS an ideal technique for probing molecular orientations of organic molecules.

Spectroscopic ellipsometry (SE) is the most commonly employed technique to determine thickness of the polymer layer in a gradient sample. SE relates thickness of the film to the change in the polarization state of light incident on the film. Specifically, SE measures two values, Δ and Ψ, which describe the polarization change in the sample and which are related to the ratio of Fresnel reflection coefficients, R_p and R_s for p- and s-polarized light, respectively. Δ and Ψ profile as a function of wavelength of light are used to obtain film thickness by performing fitting to the Δ and Ψ data. The lateral variation of polymer thickness along the gradient is measured by taking SE scans on various parts of the specimen.

While not described in this review, there are several novel experimental probes that may be applied in the near future to characterize gradient substrates with high lateral resolution (ranging from tens of square nanometers to square micrometers). These include the scanning near-field optical microscopy (SNOM) [63], imaging ellipsometry [64], imaging IR, imaging XPS [65], and X-ray photo emission electron microscopy (X-PEEM) [65]. Detailed description of these techniques is beyond the scope of this paper and the interested reader is referred to the existing literature.

3
Anchored Polymers with Grafting Density Gradients

In the following sections we will describe our recent efforts in the area of gradient polymer assemblies. For clarity, the chemical formulas of the polymers used in our experiments are shown in Fig. 11. Surface-anchored polymers with a grafting density gradient represent macromolecular systems, in which the number of polymers per unit area of the surface changes gradually as a function of the position on the surface. Figure 12 depicts the technological steps leading to the formation of surface-bound polymer assemblies with gradients in grafting density. These structures can be prepared by first generating a concentration gradient of the polymerization initiator on the surface, followed by the grafting-from polymerization [66].

3.1
Formation and Properties of the Gradient Initiator

We formed gradients of polymerization initiator on flat silica substrates using the methodology proposed by Chaudhury and Whitesides [67] (cf. Fig. 12a). Specifically, 1-trichlorosilyl-2-(m/p-chloromethylphenyl) ethane (CMPE) was

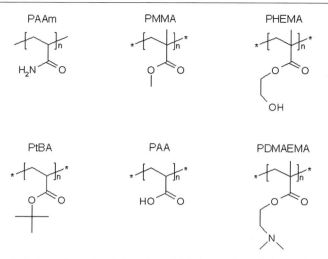

Fig. 11 Chemical formulas of poly(acryl amide) (PAAm), poly(methyl methacrylate) (PMMA), poly(2-hydroxyethyl methacrylate) (PHEMA), poly(*t*-butyl acrylate) (PtBA), and poly(acrylic acid) (PAA), and poly(dimethyl aminoethyl methacrylate) (PDMAEMA)

mixed with paraffin oil (PO) and the mixture was placed in an open container that was positioned close to an edge of a silicon wafer. As CMPE evaporated, it diffused in the vapor phase and generated a concentration gradient along the silica substrate. Upon impinging on the substrate, the CMPE molecules reacted with the substrate -OH functionalities and formed a SAM. The breadth and position of the CMPE molecular gradient can be tuned by adjusting the CMPE diffusion time and the flux of the CMPE molecules. The latter can be conveniently adjusted by varying the chlorosilane : PO ratio and the temperature of the CMPE : PO mixture. In order to minimize any physisorption of monomer and/or the polymer formed in solution on the parts of the substrate that do not contain the CMPE-SAM, we backfilled the unexposed regions on the substrate (containing unreacted -OH functionalities) with *n*-octyltrichlorosilane, (OTS). After the OTS-SAM deposition, any physisorbed CMPE and OTS molecules were removed by thoroughly washing the substrates with warm deionized (DI) water (75 °C, resistivity $> 16\,\mathrm{M\Omega \cdot m}$) for several minutes.

Near-edge X-ray absorption fine structure (NEXAFS) spectroscopy was used to provide detailed chemical and structural information about the SAMs on the substrate [60, 68]. The NEXAFS spectra were collected in the partial electron yield (PEY) mode at the normal ($\theta = 90°$), grazing ($\theta = 20°$), and so-called "magic" angle ($\theta = 55°$) incidence geometries, where θ is the angle between the sample normal and the polarization vector of the X-ray beam. In Fig. 13 we plot the carbon K-edge PEY NEXAFS spectra taken from the CMPE-SAM (top) and OTS-SAM (bottom) samples. The NEXAFS spectra col-

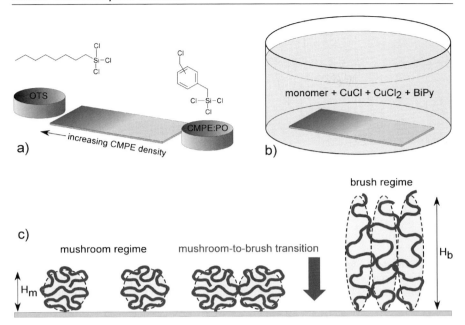

Fig. 12 Methods of preparing surface-grafted polymer assemblies with gradients in grafting density. **a** ATRP initiator gradient on a solid substrate is formed by mixing 1-trichlorosilyl-2-(m-p-chloromethylphenyl) ethane (CMPE) with paraffin oil (PO) and placing the mixture in an open container heated at 88 °C close to an edge of a silicon wafer. As CMPE evaporates, it diffuses in the vapor phase and generates a concentration gradient along the silica substrate. Upon impinging on the substrate, the CMPE molecules react with the substrate –OH functionalities and form a self-assembled monolayer (SAM). In order to minimize any physisorption of monomer and/or the polymer formed in solution on the parts of the substrate that do not contain the CMPE-SAM, the unexposed regions on the substrate containing unreacted –OH functionalities are backfilled with n-octyl trichlorosilane, (OTS). After the OTS-SAM deposition, any physisorbed CMPE and OTS molecules are removed by thoroughly washing the substrates with warm deionized water (75 °C, > 16 MΩ cm) for several minutes. **b** Surface-grafted polymer assemblies are formed on the substrates by using grafting-from ATRP. **c** Schematic illustrating polymer conformations in the mushroom (height H_m) and brush (height H_b) regimes and the mushroom-to-brush transition

lected at the magic angle were indistinguishable from those recorded at the normal and grazing incidence geometries, revealing that the CMPE-SAMs are not oriented, rather they formed a "liquid-like" structure. This observation is in accord with recent studies from the Chaudhury and Allara groups who studied the transition between the liquid-like and "semi-crystalline-like" structures in hydrocarbon SAMs [69, 70]. The NEXAFS spectra in Fig. 13 both contain peaks at 286.0 and 288.5 eV that correspond to the 1s → σ^* transition for the C – H and C – C bonds, respectively. In addition, the spectrum of CMPE also exhibits a very strong signal at 284.2 eV, which can be attributed

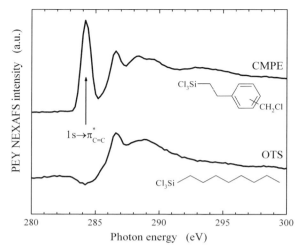

Fig. 13 Carbon K-edge partial electron yield (PEY) NEXAFS spectra collected from the CMPE-SAM (*top*) and OTS-SAM (*bottom*). The *arrow* marks the position of the 1s → π* transition for phenyl C = C, present only in the CMPE-SAM sample. (Reproduced with permission from [76])

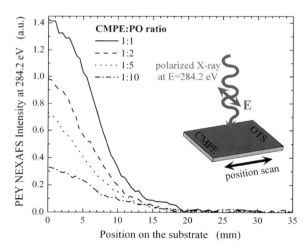

Fig. 14 Partial electron yield (PEY) NEXAFS intensity measured at E = 284.2 eV as a function of the position on the substrates containing the initiator gradients made of CMPE:PO mixtures (w/w) 1 : 1 (*solid line*), 1 : 2 (*dashed line*), 1 : 5 (*dotted line*), and 1 : 10 (*dash-dotted line*)

to the 1s → π* transition for phenyl C = C [60]. The latter signal can thus be used as an unambiguous signature of CMPE in the sample. With the X-ray monochromator set to 284.2 eV, we collected the PEY NEXAFS signal by rastering the X-ray beam across the gradient. The lines in Fig. 14 depict the variation of the PEY NEXAFS intensity measured at 284.2 eV across the gradient samples prepared by diffusing CMPE for 2 min from mixtures with various CMPE : PO ratios equal to 1 : 1 (solid line), 1 : 2 (dashed line), 1 : 5 (dotted line), and 1 : 10 (dash-dotted line). For clarity, we refer to such substrates as S1, S2, S5, and S10, respectively. The data in Fig. 14 reveal that the PEY NEXAFS intensity from the C = C phenyl bond, and thus the concentration of CMPE in the sample, decreases as one moves from the CMPE side of the sample towards the OTS-SAM; the functional form closely resembles that of a diffusion-like profile. Experiments using spectroscopic ellipsometry (SE) confirmed that only a single monolayer was formed along the substrate.

3.2
Grafting-From on the Gradient Initiator Surfaces

The polymerization of poly(acryl amide) (PAAm) was performed by atom transfer radial polymerization (ATRP), as described earlier [71–73]. In addition, PAAm brushes were grown on silica gels using the procedure outlined in [72]. The PAAm polymers were grown and purified using the same conditions as were used for growing polymer on flat silica wafers. The PAAm chains were then cleaved from the silica support and their molecular weight was measured using size exclusion chromatography (M_w = 17 kDa, polydispersity index = 1.7).

3.3
Properties of Grafted Polymer Layers

SE was used to measure the thickness of the dry polymer film, h, as a function of the position on the substrate. In Fig. 15 we plot the values of h for samples prepared on the (a) S1 (open squares), (b) S2 (open circles), and (c) S5 (open triangles) substrates. From Fig. 15, h decreases gradually as one moves across the substrate starting at the CMPE edge. Note the agreement between the variation of h and the concentration profiles of the CMPE initiator (solid lines). Because the polymers grafted on the substrate all have roughly the same degree of polymerization (see discussion below), the variation of the polymer film thickness can be attributed to the difference in the density of the CMPE grafting points on the substrate. The polymer grafting density can be calculated from $\sigma = h\rho N_A/M_n$, where ρ is the density of PAAm (= 1.302 g/cm^3).

The substrates with the grafted PAAm were placed into a solution cell that was filled with DI water (pH = 7), a good solvent for PAAm, and incubated for at least 5 h. The thickness of PAAm grafted polymer in DI water (= "wet thickness"), H, was measured using SE. The data in Fig. 15 indicate that for all samples H decreases as one traverses across the substrate starting at the

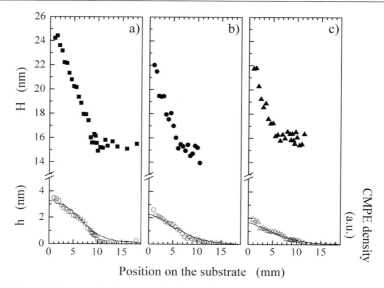

Fig. 15 Dry thickness, *h*, (*open symbols*) and wet thickness, *H*, (*closed symbols*) of poly(acryl amide) (PAAm) and the 1-trichlorosilyl-2-(m-p-chloromethylphenyl) ethane (CMPE) concentration (*solid line*) as a function of the position on the substrate for samples prepared on substrates containing the initiator gradients made of CMPE:PO mixtures (w/w) **a** 1:1 (□, ■), **b** 1:2 (○, ●), and **c** 1:5 (△, ▲). (Reproduced with permission from [76])

CMPE side. The maximum brush height at the CMPE edge of the sample decreases with decreasing CMPE concentration on the substrate (S1 > S2 > S5).

In Fig. 16 we plot the wet polymer thickness as a function of the PAAm grafting density on the substrate for all three samples, S1, S2, and S5. The data in Fig. 16 reveal that at low σ, H is independent of the grafting density. Hence the chains are in the mushroom regime. At higher polymer grafting densities, H increases with increasing σ, indicating the brush behavior. The crossover between the two regimes occurs at $\sigma \approx 0.065$ nm^{-2}. By fitting the data in the brush regime to $H \sim N\sigma^n$ we obtain n equal to 0.37 ± 0.04 (S1), 0.39 ± 0.05 (S2), and 0.40 ± 0.06 (S5). We note that n obtained by fitting the experimental data is slightly higher than the predicted value of $n = 1/3$; this observation is in agreement with recent reports [74]. A remark has to be made about the possible variation of the chain length with grafting density. Jones and coworkers recently reported on studies of grafting from polymerization of poly(methyl methacrylate) using ATRP from substrates having various surface densities of the polymerization initiator, ω-mercaptoundecyl bromoisobutyrate [75]. Their study revealed that the grafting density of the polymer depends on the grafting density of the initiator. However, based on the data presented in [75] it is difficult to discern whether the kinetics of the polymerization also depends on the grafting density of the initiator. Cur-

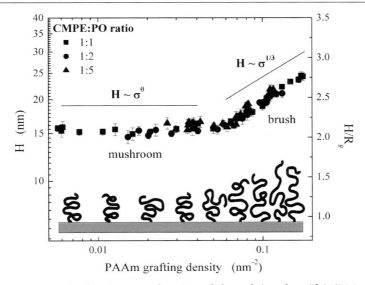

Fig. 16 Wet thickness of PAAm as a function of the poly(acryl amide) (PAAm) grafting density for samples prepared on substrates containing the initiator gradients made of CMPE:PO mixtures (w/w) 1:1 (■), 1:2 (●), 1:5 (▲). The *inset* shows a cartoon illustrating the polymer behavior. (Reproduced with permission from [76])

rently we have no means of measuring the molecular weight of the grafted brushes directly on the gradient substrate. While we cannot exclude the possibility that the length PAAm chains polymerized on the various parts of the molecular gradient substrate varies with σ, we note that the fact that the curves in Fig. 16 superimpose on a single master curve indicates that the polymers have likely very similar lengths, which is not surprising for the rather short anchored polymers synthesized in this work.

In addition to the measurement of the wet brush thickness, we have also performed wettability experiments as a function of the PAAm grafting density on the substrate [76]. Our aim was to corroborate the ellipsometric data and provide more insight into the polymer packing in the surface grafting density gradient. In Fig. 17 we plot the dry PAAm thickness, h, (closed symbols) and the contact angles of DI water, θ_{DIW}, (open symbols) as a function of the position on the substrate for samples prepared on the S1 (squares) and S5 (triangles) substrates. In both samples, the dry thickness of PAAm decreases gradually as one moves across the substrate starting at the CMPE edge. The θ_{DIW} values increase as one traverses across the substrate starting at the CMPE side. The increase in θ_{DIW} is not monotonous, it follows a "double S"-type shape. While the double S-type dependence of θ_{DIW} on the position on the sample is detected in both S1 and S5 samples, there are differences in the plateau values. Specifically, while for the S1 sample, the three plateaus are located at $\theta_{DIW} \approx 40°$, $\approx 83°$, and $\approx 100°$ the corresponding values for the S5 sample are $\theta_{DIW} \approx 47°$, $\approx 70°$,

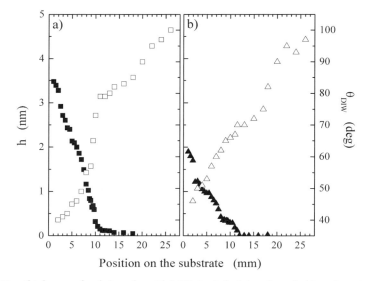

Fig. 17 Dry thickness of poly(acryl amide) (PAAm), *h*, (*closed symbols*) and contact angle of DI water, θ_{DIW}, (*open symbols*) as a function of the position on the substrate for samples prepared on substrates containing the initiator gradients made of CMPE:PO mixtures (w/w) 1:1 **a** (□, ■) and **b** 1:5 (△, ▲). The contact angle data have an error better than ±1.5 deg and ±1 mm on the vertical and horizontal scales, respectively. (Reproduced with permission from [76])

and ≈ 97°. On the basis of the dry thickness data and our previous discussion, the three plateaus in the contact angle behavior can be attributed to the wetting characteristics inside the brush, mushroom, and OTS (no PAAm) regions. At distances far away from the CMPE edge, where the θ_{DIW} values are high, there is no grafted PAAm on the sample. The contact angle experiments detect the presence of the OTS monolayer. By moving closer towards the CMPE edge, the contact angles decrease by ≈ 20–30° indicating that some polymers are present on the substrate. However, their grafting densities are low so that the probing liquid can penetrate between the grafted chains; the measured contact angles represent a weighted average between the PAAm and OTS. Upon approaching the mushroom-to-brush transition region, the contact angle further decreases. The decrease is steeper for PAAm on the S1 substrate and more gradual for the S5 sample, indicating that the density of PAAm increases more rapidly in the former case. The contact angles in the lowest plateau are $\theta_{DIW} \approx 40°$ and ≈ 47° for samples S1 and S5, respectively. In independent experiments, we have established that the θ_{DIW} of a pure PAAm is ≈ 35–38° [77]. Because in both cases the PAAm polymers grafted on the substrate have roughly the same degree of polymerization, the variation of the polymer film thickness can be attributed to the difference in the density of the CMPE grafting points on the substrate. Specifically, close to the CMPE edge, the PAAm macromolecules form a dense brush on the S1 substrate and a "semi-dense" brush on the S5 substrate.

The previous discussion revealed that θ_{DIW} depends on the grafting density of the PAAm chains on the substrate. Earlier we have shown that the wet thickness of PAAm prepared on substrates with various CMPE concentrations can be collapsed on a single master curve when plotted as H vs. σ. One would also expect that the wettabilities of the substrates plotted versus the PAAm grafting density should exhibit similar universal behavior. In Fig. 18 we plot the negative cosine of θ_{DIW} as a function of the grafting density of PAAm on substrates S1 (squares) and S5 (triangles). As anticipated, the data collapses on a single master curve. A close inspection of the results present in Fig. 18 shows that the data can be divided into three distinct regions. For $\sigma > 0.1$ nm^{-2}, the chains are expected to be in a brush regime—the wettabilities are close to the pure PAAm ($-\cos(\theta_{DIW}) \approx -0.79$). For $\sigma < 0.02$ nm^{-2} the PAAm chains form mushroom conformations on the substrate. In this regime, the wettabilities change slightly because the distance between the chains also changes, although they are already loosely separated on the substrate. At grafting densities 0.02 nm$^{-2} < \sigma < 0.1$ nm^{-2}, the slope of $-\cos(\theta_{DIW})$ changes rather rapidly. The data in Fig. 18 show that the position of the mushroom-to-brush crossover determined using the wettability approach is in accord with the ellipsometric measurements (the transition location was established to be at $\sigma \approx 0.065$ nm^{-2}). However, in the former case, the transition region extends over almost one order of magnitude in σ, which is

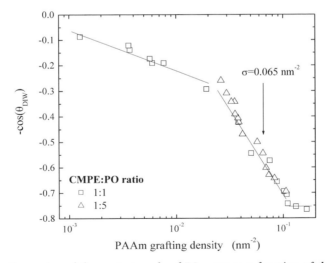

Fig. 18 Negative cosine of the contact angle of DI water as a function of the poly(acryl amide) (PAAm) grafting density on the substrate for samples prepared on substrates containing the initiator gradients made of CMPE:PO mixtures (w/w) 1:1 (□) and 1:5 (△). The *lines* are meant to guide the eyes. (Reproduced with permission from [76])

broader, as expected [14, 78–81], than the transition region predicted by the H vs. σ data. We speculate that the small difference between the widths of the mushroom-to-brush region inferred from both types of experiments is likely associated with the inaccuracy in H, which was obtained indirectly by the model fitting of the SE data.

4
Investigation of the Behavior of End-Tethered Weak Polyelectrolytes in Ionic Solutions Using Grafting Density Gradients

In the previous section, we demonstrated how the grafting density gradient of neutral polymers could be used to gain a basic understanding of the scaling properties of polymer brushes. While neutral polymer brushes constitute an important category of grafted polymers, many of the industrially relevant polymers are charged; their conformational characteristics are different from those of neutral polymers. Comprehending and tailoring the behavior of charged molecules (i.e., polyelectrolytes) at surfaces and interfaces is important in designing and utilizing novel applications (e.g., pH-controlled flow through polymeric micro-membranes), many of which cannot be fabricated using any other set up. Study of polyelectrolytes is also necessitated by the ubiquity of charged macromolecules found in biological systems. The behavior of polyelectrolyte brushes is quite complex because it is governed by a whole battery of parameters, which include both thermodynamic and electrostatic interactions. Specifically, in addition to the parameters governing the performance of neutral polymer brushes, i.e., polymer molecular weight (or equivalently the degree of polymerization, N), brush grafting density, σ, and the solvent quality (characterized by the Flory–Huggins interaction parameter, χ), the properties of polyelectrolyte brushes depend strongly on the degree of dissociation of the backbone charges (or degree of dissociation), α, counterion volume fraction in the polyelectrolyte solution, counterion valency, q, external salt concentration, $\rho_{s,\infty}$, and in some cases pH of the solution. Gradients, by virtue of their combinatorial nature, are best suited to investigate the behavior of such complex systems. In the following section we briefly outline several scaling approaches describing the behavior of charged surface-bound polymers. Following that, we describe the results of our recent experiments involving grafting density gradients of weak polyelectrolyte brushes made of poly(acrylic acid) (PAA).

4.1
Theoretical Models of Surface-Bound Polyelectrolytes

Various theoretical approaches have been utilized to describe the performance of charged macromolecules at interfaces. In particular, scaling theories pioneered by Pincus [82], Zhulina and coworkers [83, 84] and the Wageningen group [18, 85–88] laid the foundation of our current understanding of polyelectrolyte brush behavior. These studies revealed that several different regimes of polyelectrolyte brushes could be identified depending on the concentration of the external salt in solution. Presence of salt in the solution has different effects on the polyelectrolyte depending on whether the polymer is a strong or weak polyelectrolyte. Strong (or "quenched") polyelectrolytes have fixed α; their properties thus do not depend on the pH of the solution. On the other hand, in weak (or "annealed") polyelectrolytes, α depends on pH. At high $\rho_{s,\infty}$ the salt concentration inside and outside the brush is about the same and the electrostatic interactions are largely screened. Under such conditions the polyelectrolyte brush behaves exactly as a neutral brush (NB) and $H/N \approx (v\sigma)^{1/3}$, where H is the height of the polymer brush and $v(= 0.5 - \chi)$ is the excluded volume parameter. When the external salt concentration decreases, there is an unbalance in the ion concentration inside and outside the brush because the polymer charge density inside the brush ($\alpha\varphi$, where φ is the polymer volume fraction) is no longer negligible with respect to $\rho_{s,\infty}$. The system enters the so-called salted brush (SB) regime. In the SB regime, H/N scales as $H/N \approx (v_e\sigma)^{1/3}$, where $v_e(= \alpha^2/\rho_{s,\infty})$ is the electrostatic excluded volume parameter. Because of the electrostatic interactions inside the brush, a salted brush is more extended than a neutral one. As shown schematically in Fig. 19, the brush expansion increases with decreasing $\rho_{s,\infty}$. If the external salt concentration is further decreased such that $\rho_{s,\infty} \ll \alpha\varphi$

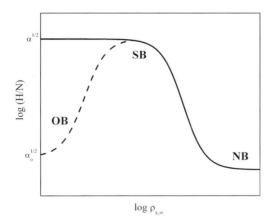

Fig. 19 Dependence of the brush thickness reduced by the number of polymer repeat units for monovalent co-ions, H/N, on the concentration of the external salt, φ_s, for strong (*solid line*) and weak (*dashed line*) polyelectrolyte brushes in neutral brush (NB), salted brush (SB), and osmotic brush (OB) regimes. α and α_o denote the bulk and "internal" (for weak polyelectrolyte brushes only) degree of dissociation, respectively (Reproduced with permission from [89])

the co-ions are effectively expelled from the brush and $H/N \approx \alpha^{1/2}$. In this so-called osmotic brush (OB) regime, a limiting brush thickness is reached, which is independent of $\rho_{s,\infty}$ and σ.

The behavior of weak polyelectrolyte brushes is different from that of strong polyelectrolyte brushes. Here the number of the backbone charges is not fixed. Specifically, α depends on the proton concentration in the polymer solution, $[H^+] = 10^{-pH}$, and is given by $\alpha/(1 - \alpha) = K/[H^+]$, where K is the dissociation constant. When there is an excess of salt, as in the NB and SB regimes, $[H^+]$ inside and outside the brush is approximately equal and the internal degree of dissociation is the same as that in the bulk solution. Hence, the scaling for H/N in the NB and SB regimes is the same as in the case of strong polyelectrolyte brushes:

$$ H \sim N\sigma^{1/3} \left(\frac{\alpha^2}{\rho_{s,\infty}} \right)^{1/3} . \tag{1} $$

When the system enters the OB regime, a significant electric potential difference develops between the brush and the solution in the bulk. In addition, $[H^+]$ inside the brush is considerably higher. As a consequence, a portion of the brush charges associate with protons and $\alpha_0/(1 - \alpha_0) = K/(\sigma\alpha_0^{1/2})$, where α_0 is the internal degree of dissociation. This value of α_0 may be much smaller than the value in the bulk (α); the weak groups respond to the unfavorable electrostatic condition in OB by discharging themselves. The brush height in the OB regime is predicted to scale as [83]:

$$ H \sim N\sigma^{-1/3} \left(\frac{\alpha}{1 - \alpha} \right)^{1/3} ([H^+] + \rho_{s,\infty})^{1/2} . \tag{2} $$

Such a response is impossible for strong brushes, which have a fixed α. Figure 19 illustrates the different behavior of weak polyelectrolyte brushes in the OB regime (dotted curve). Because of the discharging process ($\alpha_0 < \alpha$), a weak brush in the OB regime is less expanded than the strong brush. As a result, H/N passes through a maximum as a function of $\rho_{s,\infty}$, being small for both high and small $\rho_{s,\infty}$. The unusual feature that at low $\rho_{s,\infty}$ the brush contracts with decreasing $\rho_{s,\infty}$ is a typical property of weak groups, which can respond to a change in the local environment.

While successful in providing a qualitative picture of the behavior in charged surface-tethered systems, these scaling approaches often do not provide quantitative information about the various characteristics of charged polymer systems. This is understandable, as some of the requirements imposed on these scaling approaches—such as, mean-field behavior, charge and density homogeneity in the plane parallel to the substrate, just to name a few—are not typically fulfilled in real experimental conditions. Thus, in order to quantitatively model the behavior of charged brushes, a more detailed and more complicated theoretical treatment is needed. One such approach is a generalized form of a single chain molecular theory, developed by

Szleifer and coworkers. Recently, Wu et al. applied this theory to successfully model the behavior in charged polymer brush systems [89]. More importantly, this theory enables the estimation of system parameters that are not easy (or even possible) to measure experimentally, such as the local dissociation inside the brush and the average charge of the polymer as a function of the distance from the surface. We will not provide any detailed information about the theory here. The interested reader is referred to the original publication and the references cited therein [89].

4.2
Formation of the Grafting Density Gradient of PAA Brushes on Flat Substrates

Several groups investigated the behavior of weak polyelectrolytes anchored at surfaces. For example, Currie and coworkers prepared poly(acrylic acid) (PAA) brush samples with three different grafting densities and studied their solution behavior at three low pH values as a function of the solution ionic strength [90, 91]. They found that the PAA wet brush thickness is a nonmonotonous function of the ionic strength at a given pH and grafting density. The extent of swelling of the brush increased with increasing pH and grafting density. Although the nonmonotonous behavior agreed qualitatively with theoretical predictions, the mean-field power law for the OB regime at a given pH and σ, cf. Eq. 2, was not observed. We studied the conformational behavior of grafted polyelectrolyte chains by employing grafting density gradients of PAA. The aim of the work was to characterize PAA interfacial properties as a function of σ, pH, and ionic strength (IS).

Previous studies have revealed that acrylic-based polymers are difficult to prepare by ATRP because of the interaction of the carboxylic acid functionalities with the ATRP catalyst [41]. Hence, in order to form a surface-anchored PAA with a grafting density gradient, we first used the method described in Sect. 3 to synthesize a gradient of poly(t-butyl acrylate) (PtBA) and then converted PtBA into PAA by acid wash hydrolysis of PtBA [92, 93]. The only difference between the method described in Sect. 3 and the one adapted to synthesize PtBA was that OTS was first evaporated over a silicon wafer so as to generate a concentration gradient of OTS and the substrate was then back-filled with the polymerization initiator, thus forming a gradient of initiator molecules in the reverse direction. This modification was necessary because the vapor pressure of the ATRP initiator used in this study, (11-(2-Bromo-2-methyl)propionyloxy) undecyltrichlorosilane (BMPUS), was not sufficiently high and its evaporation was not feasible without possible thermal damage to the molecule. The PtBA was converted into PAA by treating the specimens with a mixture of 1,4-dioxane and HCl and the conversion was verified by Fourier transform infrared spectroscopy (FTIR) [89]. To estimate the molecular weight of polymers grown on a surface, solution polymerization of tBA with methyl 2-bromopropionate as the initiator was also carried out. The experimental conditions for solution polymerization were identical to those used for surface-initiated polymerization. We note the polymerization rate in the solution is likely faster than that on the flat substrate [37]. Nevertheless, the molecular weight of the solution-based polymer provided a useful estimate for further analysis.

SE was used to measure the dry thickness of both the PtBA and PAA gradient samples, which were polymerized for 10 h and hydrolyzed in a HCl/dioxane bath for 5 h. In Fig. 20, we plot the dry thickness of PtBA (solid symbols) and PAA (open symbols) as a function of the position on the substrate. The data in Fig. 20 reveal that the thickness of both PtBA and PAA increases as one moves from the OTS side (small number on the abscissa) of the sample towards the initiator-covered side (large numbers on the abscissa); in both cases the functional form closely resembles that of a backward diffusion-like profile. Assuming that all chains of both PtBA and PAA have the same degree of polymerization along the substrate (i.e., the polymerization rate was independent of the grafting density of the initiator on the substrate [66, 76]), the increase of the polymer dry thickness can be attributed to the increase of the polymer grafting density on the substrate. The solid line in Fig. 20 depicts the variation of the PEY NEXAFS intensity corresponding to the C = O bond across the PAA gradient. The NEXAFS results confirm that the C = O intensity, and thus the amount of PAA on the surface, increases as one moves along the gradient. Moreover, the NEXAFS data are in good agreement with the ellipsometric thickness of PAA. More discussion on the hydrolysis of PtBA into PAA can be found in [89].

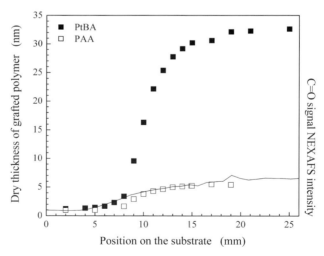

Fig. 20 Dry thickness of poly(t-butyl acrylate) (PtBA, ■) and poly(acrylic acid) (PAA, □) as a function of the position on the substrate. The *solid line* represents the PEY NEXAFS intensity measured at E = 531 eV on the PAA sample as a function of the position on the substrate. (Reproduced with permission from [89])

4.3
Solution Properties of PAA

In order to study the solution properties of the surface-grafted PAA, we measured the wet thickness of PAA in aqueous solution by placing the samples in a custom-designed solution cell, incubating them for a desired period of time (typically > 5 h) and performing the experiments with SE. Aqueous solutions having different pH values (4, 5.8, and 10) and a series of ionic strengths for each pH were used. By measuring the wet thickness of PAA along the gradient at different solution conditions, we obtained the wet thickness of the grafted polymer layer as a function of the PAA grafting density, pH, and ionic strength. The SE experiments were carried out on PAA with M_n = 4.8 kDa. The wet PAA thickness was evaluated using a graded effective medium approximation model based on linear combination of the optical constants of the DI water and PAA [76].

4.3.1
Dependence of Wet PAA Thickness on Ionic Strength

In Fig. 21 we plot the dependence of the PAA wet thickness (H) on the solution ionic strength (IS) at pH equal to (a) 4, (b) 5.8, and (c) 10 for three different grafting densities. The squares, circles, and triangles denote grafting density values that are approximately equal for all three samples. Since only NaCl, HCl, and NaOH were used to change the solution ionic strength, the salt concentration ($\rho_{s,\infty}$) in this case is equal to the solution ionic strength (IS). The data in Fig. 21 reveal that H depends on IS in a nonmonotonous fashion. Specifically, as IS increases, H increases before reaching a maximum at a certain value and then starts to decrease. This behavior, observed for all pH values at all values of σ, is in accord with the theoretically predicted trends [87] that divide the H vs. IS dependence into the osmotic brush (OB) and the salted brush (SB) regimes (cf. Fig. 19). The ionic strength, at which the transition between the OB and SB regimes occurs (IS_{max}), is related to σ and pH. At pH = 4, IS_{max} is nearly constant regardless of the σ (cf. Fig. 21a). At pH = 5.8, IS_{max} remains small at low grafting densities and increases slightly to 0.25 with increasing σ (cf. Fig. 21b). At pH = 10, IS_{max} shifts significantly (cf. Fig. 21c). Specifically, while for low σ (≈ 0.0381 nm^{-2}, not shown) $IS_{max} \approx 0.25$, at high σ (≈ 0.863 nm^{-2}) $IS_{max} \approx 1$.

Overall, two general trends can be deduced from the data in Fig. 21. First, with increasing grafting density, the height of the polymer brush increases. Second, at a given value of σ, the PAA swelling increases as the solution pH value increases. The latter behavior is associated with the electrostatic charging inside the PAA brush, which leads to the increase of the intermolecular repulsions and subsequent brush height increase.

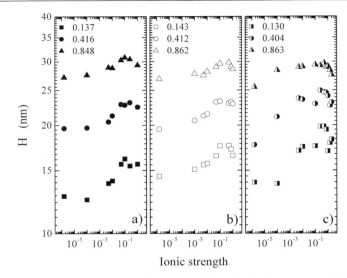

Fig. 21 Wet thickness of PAA (H) as a function of the solution ionic strength (IS) at **a** pH
= 4, **b** pH = 5.8, and **c** pH = 10. The symbols represent different grafting densities of PAA
in chains/nm^2. (Reproduced with permission from [89])

4.3.2
Dependence of Wet PAA Thickness on Grafting Density

Earlier we discussed that weak polyelectrolyte wet thickness has a different dependence on the grafting density for polymers in the SB or OB
regimes [83]. In this section we will present the results of the experimentally
measured H for each regime separately. Since there is no significant variation in the data for different pH values and molecular weights, we will use the
results collected at pH = 5.8 and PAA with M_m = 4.8 kDa.

Salted Brush (SB) Regime

In Fig. 22b we plot H as a function of the PAA grafting density in the SB
regime. The various symbols denote data collected at IS ranging from 0.1 to
0.75. At high polymer grafting densities ($\sigma > 0.1$ nm^{-2}), H increases with increasing σ. This is a typical behavior for the brush conformations. The transition from the brush regime to the mushroom regime occurs at $\sigma \approx 0.08$ nm^{-2}.
The slope for the brush regime is found to range from 0.29 to 0.31, in good
agreement with the theoretically predicted value of 1/3. With increasing IS,
H decreases and the slope in the $H \sim \sigma^n$ dependence increases. The decrease
in polymer swelling is largely due to the screening of the electrostatic interactions by the counter ions inside the polymer brush. The increase in the slope
suggests that the solution ions move more easily inside the grafted polymer

at lower grafting density. With increasing σ, the transport of ions inside the densely packed polymers becomes harder. As a consequence, the screening effects weaken.

Osmotic Brush (OB) Regime

In Fig. 22a, we plot H as a function of σ for IS ranging from 1.56×10^{-6} to 0.1 M. At these IS values the system is in the OB regime. Prior theoretical work predicted that in this regime the wet thickness of a polymer brush should *decrease* with the grafting density as $H \sim \sigma^{-1/3}$ and should increase with increasing IS [83, 85]. On the basis of theoretical studies, at the transition between the OB to SB regimes (at IS_{max}), H is independent of the brush grafting density. Similar to earlier experiments by others [90, 91], we observe that this scaling relation is somehow flawed. Specifically, by fitting the data in the brush regime to $H \sim N\sigma^n$, we obtain n that ranges from 0.28 to 0.34 instead of the expected value of $-1/3$. Close inspection of the data in Fig. 22 reveals that polymer swelling increases with increasing ionic strength. Interestingly, the value of the exponent n decreases systematically as the solution IS increases. This is in contrast to the performance of PAA in the SB regime, where the value of n increased with increasing IS (cf. Fig. 22b). This behavior reveals that when a small amount of salt is added in the OB regime to polymers with a low σ, the grafted polymer swells more relative to PAA at high σ.

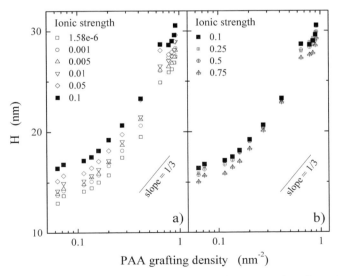

Fig. 22 Wet thickness at pH = 5.8 for PAA ($M_n = 4.8\,$kDa) as a function of the grafting density and ionic strength of the aqueous solution in the **a** OB regime and **b** the SB regime. The *symbols* represent different IS values. (Reproduced with permission from [89])

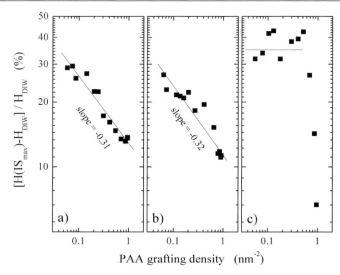

Fig. 23 Degree of swelling $[H(IS_{max}) - H_{DIW}]/H_{DIW}$ for PAA (M_n = 4.8 kDa) as a function of the PAA grafting density in the OB regime at pH equal to **a** 4, **b** 5.8, and **c** 10. (Reproduced with permission from [89])

In order to quantify this behavior, we define a degree of swelling ($[DS]$) of a grafted polymer as:

$$DS = \frac{H(IS) - H_{DIW}}{H_{DIW}} \times 100\% , \tag{3}$$

where $H(IS)$ and H_{DIW} are the PAA thicknesses evaluated at a given IS and in pure water ($IS \to 0$), respectively. In Fig. 23, we plot the degree of swelling at IS_{max} as a function of the PAA grafting density at different pH values for polymers in the OB regime. By fitting the data to $[DS] \sim \sigma^n$, we find n to be very close to – 1/3 in the OB regime for pH = 4 and 5.8, and 0 for the pH = 10 data. At low pH, PAA behaves as a weak polyelectrolyte, the degree of swelling changes with the σ. At pH = 10, almost all charges along the polymer backbone are activated and present at the backbone. As a consequence, the polymer behavior closely resembles that of a strong polyelectrolyte, whose degree of expansion is independent of the polymer grafting density.

5
Anchored Polymers with Molecular Weight Gradients

In addition to the polymer grafting density, polymer molecular weight is another important molecular parameter that profoundly influences the properties of surface-anchored polymers. As discussed earlier, the thickness of the grafted polymer layer is proportional to the degree of polymerization of the anchored polymer. For some applications to be discussed subsequently in this review, it would be convenient to have samples with anchored polymers having variable degrees of polymerization.

We have recently designed a method leading to the preparation of surface-anchored polymers with a variable degree of polymerization [94]. The samples are prepared in a polymerization chamber with a vertically positioned sample holder. Tubing, attached at the bottom of the chamber, is connected to a micropump, which controls the flow rate of removal of the solution from the chamber (cf. Fig. 24). The polymer brush formation proceeds as follows. The chamber is initially loaded with a solution comprising a monomer, bipyridine, $CuCl_2$, and the solvent. The chamber is purged with nitrogen for a couple of minutes in order to remove any oxygen present. CuCl is added and the silicon wafer, covered with a chemisorbed ATRP initiator, is lowered into the solution. The polymerization proceeds following the standard ATRP scheme [41, 42, 53, 54]. During the reaction, the micropump removes the solution from the chamber causing a steady decrease in the vertical position of the 3-phase (substrate/solution/inert) contact line. It has to be noted that in principle, there may be a thin layer of liquid present on the polymer substrate immediately after it is removed from the polymerization solution. The thickness of the liquid film dragged upon solvent removal depends on the

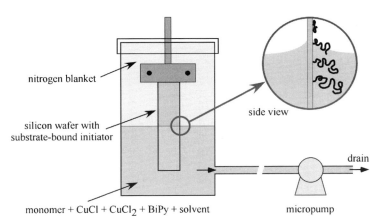

Fig. 24 Schematic of the apparatus for creating surface-grafted polymer assemblies with a gradient in molecular weight. The chamber is loaded with a solution comprising a monomer, bipyridine, $CuCl_2$, and the solvent. The chamber is then purged with nitrogen in order to remove oxygen. CuCl is added and the silicon wafer, covered with a chemisorbed ATRP initiator, is lowered into the solution. During the reaction, the micropump removes the solution from the chamber causing a steady decrease in the vertical position of the 3-phase (substrate/solution/inert) contact line. The profile (including the "steepness") of the polymer brush length gradient on the substrate can be controlled by varying the removal rate. (Reproduced with permission from [94])

nature of the polymer. Strongly and moderately hydrophobic polymers, such as poly(methyl methacrylate) (PMMA) or poly(2-hydroxylethyl methacrylate) (PHEMA), do not form such a film and water/methanol mixture simply beads down the substrate. For hydrophilic polymers, a thin liquid film may appear upon removal of the reaction mixture. This film, however, subsequently flows down the substrate due to gravity. This motion of the film works in the same direction as the removal of reaction mixture, i.e. top portion of the exposed substrate remains in contact with the liquid film for a shorter duration than the bottom portion of the exposed substrate. On complete drainage of the reaction mixture, the sample is removed, washed, and sonicated thoroughly with de-ionized water and blow-dried with nitrogen.

Poly(methyl methacrylate) with a variable degree of polymerization anchored to silica surfaces was synthesized following the room temperature ATRP polymerization scheme described earlier [45, 46]. In the main part of Fig. 25 we plot the variation of the PMMA brush thickness after drying (measured by SE) as a function of the position on the substrate. Thickness increases continuously from one end of the substrate to the other. Since the density of polymerization initiators is (estimated to be ≈ 0.5 chains/nm^2) uniform on the substrate, we ascribe the observed change in thickness to different lengths of polymer chains grown at various positions.

The combinatorial design of our system is conveniently suited for kinetic studies of ATRP. The gradient apparatus allows for complete probing of the

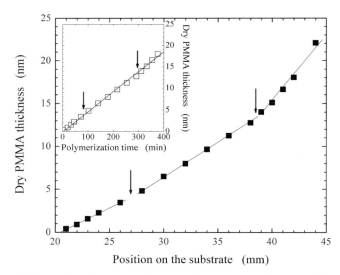

Fig. 25 Dry poly(methyl methacrylate) (PMMA) thickness as a function of the position on the silica substrate. The *inset* shows the dry PMMA thickness as a function of the polymerization time. The *arrows* mark instances where the drain speed of the polymerization solution from the chamber was decreased relative to the previous drain speed. The *lines* are meant to guide the eye. (Reproduced with permission from [94])

anchored polymer properties and studying the polymerization kinetics in confined geometries. For example, the thickness profile in Fig. 25 is controlled by the removal rate of polymerization solution. The arrows in Fig. 25 mark the instances where the exhaust speed was reduced two times during the polymerization. This results in three separate regions having different slopes in the thickness profile along the substrate. The inset depicts the variation of thickness but as a function of the polymerization time. The data in the inset illustrate that the thickness increases linearly with polymerization time, in accord with previous reports [45, 54]. Thus, the polymerization rate as a function of time was not affected but the "steepness" of the gradient was controlled by varying the drainage rate. It is worth noting that using the gradient apparatus, entire polymerization kinetics can be condensed onto a single substrate thus obviating the need for the preparation of numerous samples under similar conditions usually required for such study. With the gradient apparatus, it is quite straightforward to distinguish between "ideal" and "nonideal" controlled/"living" radical polymerizations. In an ideal controlled/"living" radical polymerization the number average molecular weight of growing polymers scales linearly with monomer conversion [42, 53, 54]. Any deviation from linearity can thus contribute to the nonliving character of polymer growth. For example, the inset in Fig. 25 clearly shows that PMMA polymerization is well controlled with a linear thickness profile, indicating a constant polymerization rate. However, polymerization of acryl amide (AAm) produces a nonlinear thickness profile as seen in the inset of Fig. 26. Rapid initial growth followed by abrupt leveling of thickness indicates uncontrolled growth with a PAAm/bipyridine/$CuCl_2$/CuCl system. Nonetheless, a linear thickness profile as a function of position on the substrate can be obtained even for PAAm by adjusting the drainage rate of the polymerization mixture from the reaction chamber (cf. Fig. 26).

Our experiments revealed that the parameters of the polymer brush—notably the polymer growth rate and polydispersity—are controlled by the amount of $CuCl_2$ added to the reaction vessel (cf. Fig. 27). At higher $CuCl_2$/CuCl ratio, a greater control is obtained over the polymerization reaction, albeit at the expense of decreasing the reaction rate. This finding is not that surprising given the nature of the reaction. The key reaction in ATRP is the reversible activation–deactivation process using metal (M)/ligand (L) complexes:

$$P - X + M^{I}X/2L \underset{k_{d}}{\overset{k_{a}}{\longleftrightarrow}} P^{*} + M^{II}X_{2}/2L \tag{4}$$

where k_a and k_d are the rate constants for activation and deactivation, respectively, M is usually Cu, X is Cl or Br. The propagating radical, P^*, produced by the halogen atom transfer from $P - X$ to the $M^{I}X/2L$ complex will undergo polymerization until it is deactivated by the $M^{II}X_2/2L$ complex. The quick speed of the activation–deactivation cycles compared to the rate of polymer-

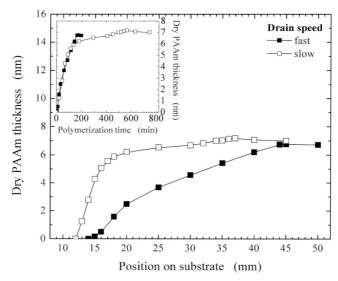

Fig. 26 Dry thickness of poly(acryl amide) as a function of the position on the silica substrate prepared by slow (□) and fast (■) removal of the polymerization solution by utilizing the method depicted in Fig. 24. The *inset* shows the dry poly(acryl amide) thickness as a function of the polymerization time. Note that both data sets collapse on a single curve at short polymerization times. Regardless of the drain speed, the brush thickness increases linearly at short polymerization times and levels off at longer polymerization times. The latter behavior is associated with premature termination of the growing polymers

ization and the low concentration of the active species (relative to the $P - X$ ones) lead to polymers with narrow polydispersities. MX_2 is usually added to the reaction mixture to regulate the reaction rate and chain polydispersity.

More insight into the ATRP polymerization on solid substrates can be obtained by evaluating the polymerization rate as a function of the $CuCl_2/CuCl$ ratio. Matyjaszewski and coworkers [54, 95] established that the rate of ATRP polymerization, R_p, is given by Eq. 5:

$$R_p = k_p \frac{k_a}{k_d} \left[P - Cl \right] \left[M \right] \frac{[CuCl]}{[CuCl_2]} , \tag{5}$$

where k_p is the rate constant for propagation, $[P - Cl]$ is the concentration of the growing ends of the grafted polymer, $[M]$ is the concentration of the free monomer in the solution, $[CuCl]$ and $[CuCl_2]$ are the concentrations of CuCl and $CuCl_2$, respectively. Recognizing that $R_p \sim dh/dt$ and lumping the terms that stay constant during the polymerization, one arrives at Eq. 6:

$$\frac{dh}{dt} \sim \left(\frac{[CuCl_2]}{[CuCl]} \right)^{-1} \tag{6}$$

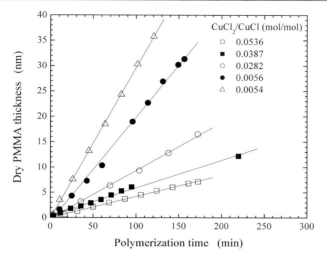

Fig. 27 Dry poly(methyl methacrylate) thickness as a function of the polymerization time for various CuCl$_2$/CuCl ratios: 0.0536 (□), 0.0387 (■), 0.0282 (○), 0.0056 (●), and 0.0054 (△). The *lines* are linear fits to the data. (Reproduced with permission from [94])

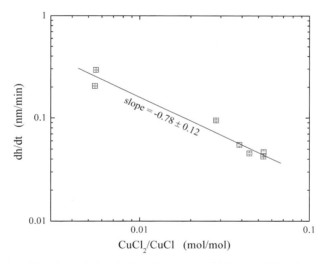

Fig. 28 Change of the dry poly(methyl methacrylate) thickness with polymerization time (dh/dt) as a function of the CuCl$_2$/CuCl ratio. The dh/dt values were obtained by fitting the experimental values of h vs. t in Fig. 27. (Reproduced with permission from [94])

Figure 28 depicts dh/dt obtained by fitting the dry thickness vs. polymerization time data (cf. Fig. 27) as a function of the CuCl$_2$/CuCl ratio. By fitting the data to a line on the logarithmic scale we obtain an exponent of $\approx -0.78 \pm 0.12$. This value is slightly lower than the exponent predicted by the scaling argument presented in Eq. 6. There are multiple factors that may influence the absolute value of the exponent. First, the slope ($= dh/dt$) is extremely sensitive to the amount of $CuCl_2$ at very low $CuCl_2$ concentrations. Second, the scaling argument in Eq. 6 is derived from the ATRP rate of polymerization that has been derived for bulk polymerization and does not take into account any constraints due to confinement effects. Third, at low $CuCl_2$ content the ATRP polymerization proceeds quite rapidly (cf. Fig. 27) and the nature of the reaction is less controlled relative to the situations at higher $CuCl_2$ concentrations.

6
Grafted Polymers with Chemical Composition Gradients

Due to its living nature, ATRP has been remarkable in its ability to produce block copolymer brushes [28, 96, 97]. We combined the gradient preparation method described in Sect. 5 with the living character of ATRP to create complex architectures that would otherwise be difficult to build [98]. For example, the brush length gradient of poly(methyl methacrylate) (PMMA), can be subsequently immersed into a solution of another monomer, say styrene, thus forming a grafted diblock copolymer PS-b-PMMA. This set up is possible because the PMMA brush acts as a living macroinitiator for the subsequent polymerization of PS. Depending on which side of the styrene length gradient is immersed first into the styrene solution, either a copolymer brush with a variable length but a constant composition (longer PMMA chains first), or a copolymer with a constant length but variable composition (shorter PMMA chains first) is formed. In the previous example, we assumed for simplicity that the rates of immersion would be adjusted such that the rates of polymerization of styrene and methacrylate would be equal.

We have succeeded in preparing the first polymer composition gradient brush anchored on the solid substrate. Specifically, we first anchored the PHEMA brush with a molecular weight gradient. We then used a PHEMA block as a macroinitiator for the polymerization of methyl methacrylate (MMA). By changing the way the PMMA block is grown on top of the PHEMA macroinitiator, PHEMA-b-PMMA copolymers with a variable PHEMA block length and either a constant (sample S1) or variable (sample S2) PMMA block length can be designed (cf. Fig. 29). In order to generate sample S1, the substrate covered with the PHEMA surface-tethered block is immersed into a solution containing MMA and removed after a fixed reaction time. The resulting copolymer (arrow C in Fig. 29) has a variable PHEMA and constant PMMA block lengths. Alternatively, PHEMA-b-PMMA can be formed that has a constant total length and a gradual composition variation along the substrate ranging from pure PHEMA to pure PMMA homopolymers (sample S2). In order to achieve this, the polymerization chamber is filled with a MMA polymerization solution. The substrate covered with the PHEMA brush hav-

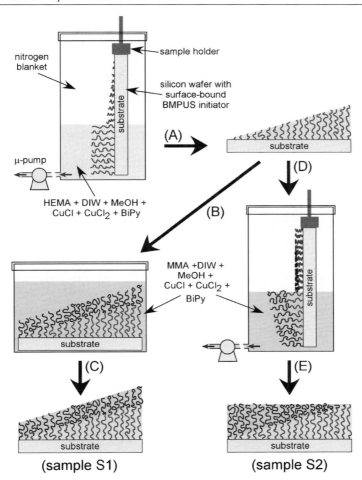

Fig. 29 Schematic illustrating the technological steps leading to the formation of surface-anchored poly(2-hydroxyethyl methacrylate)-block-poly(methyl methacrylate) (PHEMA-*b*-PMMA) assemblies with a gradual composition variation along the substrate. A custom designed apparatus is used to decorate the sample surface with a grafted PHEMA having a gradient in molecular weight (*arrow A*). While polymer immersed in the polymerization solution is fully expanded, macromolecules on the dry part of the substrate are collapsed. Surface-grafted PHEMA acts as a macroinitiator for the polymerization of the PMMA block that has either a constant molecular weight (*arrow B*) or a variable molecular weight (*arrow D*). The overall process results in PHEMA-*b*-PMMA block copolymers with a constant PMMA length and a variable total length (*arrow C*) or a gradual PMMA length and a constant total length (*arrow E*). (Reproduced with permission from [98])

ing a molecular weight gradient is lowered into the chamber with the side containing the shorter PHEMA block pointing down (arrow D in Fig. 29). The polymerization bath is pumped out from the chamber, which leads to the formation of a PMMA "reverse" molecular weight gradient on top of the PHEMA molecular weight gradient (arrow E in Fig. 29).

In Figs. 30a and 30b we plot the variation of the dry thickness of the PHEMA block (squares) and the PHEMA-*b*-PMMA copolymer (circles) in samples S1 and S2, respectively. The smaller amount of $CuCl_2$ and the slower drain rate of the HEMA polymerization solution in sample S2, resulted in faster PHEMA polymerization (and thus a thicker PHEMA layer) and a steeper PHEMA gradient. The data in Fig. 30a illustrate that the PMMA block in sample S1 has a uniform thickness regardless of the length of the PHEMA macroinitiator. In contrast, the PMMA thickness in sample S2 increases with decreasing the PHEMA length. While this behavior is expected, Fig. 30b demonstrates that the total PHEMA-*b*-PMMA thickness does not stay constant across the sample. Several parameters are expected to influence the properties of the copolymers. First, the activity of the macroinitiator (PHEMA in this case) would crucially depend on the length. Second, we may be suffering from complications in wetting at the three-phase contact line meniscus. Considering that the initiator-covered substrates are hydrophobic and that the PHEMA grown onto the substrate is slightly hy-

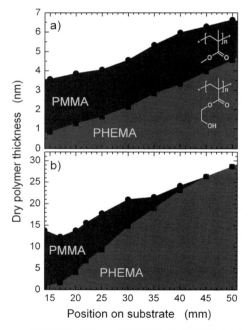

Fig. 30 Dry thickness of PHEMA (■) and PHEMA-*b*-PMMA (●) in samples **a** S1 and **b** S2. The S1 and S2 PHEMA-*b*-PMMA copolymers have been prepared following the route A → B → C and A → D → E, respectively, depicted in Fig. 29. (Reproduced with permission from [98])

drophilic, the meniscus shape and the wetting angle may initially change during the polymerization. This may explain the concave thickness profile of PHEMA in the initial stages of the polymerization. While more work is needed to fine-tune the preparation of copolymer assemblies with variable molecular weight and composition, the novel strategy exemplified here may be easily extended to create gradients of numerous other types of polymers such as hydrophilic/hydrophobic, ionomers, polar/nonpolar, and conductive/insulating polymers.

7
Investigation of Solution Behavior of Polyampholytes using Gradients

Polyampholytes are macromolecules that bear both positive and negative charges along their backbone; these charges are either permanent (strong or quenched) or can be induced by varying pH of the solution (weak or annealed). As in any other polyelectrolyte system, the behavior of polyampholytes in solution is governed by a large set of parameters, including the degree of polymerization of the polymer, the degree of dissociation (= number of charges), concentration of external salt (or the solution ionic strength), and the interaction parameters between the polymer segments and solvent molecules. In contrast to polyelectrolytes with only one type of charge, the solution conformational properties of polyampholytes depend markedly on the relative amounts of the two charges and their distribution along the backbone. Polyampholytes with a net charge are dominated by strong repulsion between like-charged monomers, which tend to extend the chain dimension and help to dissolve the chains even in solvents that are poor for the backbone (the so-called polyelectrolyte effect). The charge sequence in "neutral" polyampholytes is important. For example, while polyampholytes with an alternating sequence of the opposite charges are typically swollen, block (or blocky) copolymers may collapse and precipitate from solution [99–101].

While the bulk behavior of polyampholytes has been investigated for some time now, studies of interfacial performance of polyampholytes are still in their infancy. There are several reasons for the limited amount of experimental work; the major one being the rather complex behavior of polyampholytes at interfaces. This complexity stems from a large array of system parameters governing the interaction between the polymer and the substrate. Nearly all interfacial studies on polyampholytes reported to-date involved their adsorption on solid interfaces. For example, Jérôme and Stamm and coworkers studied the adsorption of poly(methacrylic acid)-block-poly(dimethyl aminoethyl methacrylate) (PMAA-b-PDMAEMA) from aqueous solution on silicon substrates [102, 103]. The researchers found that the amount of PMAA-b-PDMAEMA adsorbed at the solution/substrate interface depended on the solution pH. Specifically, the adsorption increased as the isoelectric point (IEP) of the polyampholyte was approached—both from the small and the large pH values and a maximum in adsorption amount was observed at a pH close to IEP. At the IEP, virtually all copolymer precipitated in solution [104] and almost no adsorption at the substrate was detected. To the best of our knowledge, to date no work has been reported on the performance of surface-tethered polyampholyte copolymers. Our group has recently initiated a project aimed at generating polymer assemblies of surface-anchored polyampholyte diblock copolymer brushes comprising poly(dimethyl aminoethyl methacrylate) (PDMAEMA) and poly(acrylic acid) (PAA) blocks and investigating their interfacial behavior. In this section we briefly outline the results of our preliminary experiments and illustrate the complexity of the interfacial behavior of these multi-ion macromolecules.

Since PDMAEMA and PAA are weak polycationic and polyanionic electrolytes respectively, the number of charged monomer along the backbone varies with pH. The bulk pK_a values, i.e., the pH at which half of the ionizable groups along the chains are activated (= "charged") of PDMAEMA and PAA are 7.5 [105] and 4.8 [106, 107], respectively [89, 108]. At pH $\ll$ 4.8, the system contains a large number of positively charged PDMAEMA segments and neutral PAA chains. Similarly, at pH $\gg$ 7.5, PAA is heavily negatively charged, while PDMAEMA contains almost no charges. In those two pH regimes one would expect to see the polymer conformation to be dominated by the charges along the backbone. Interesting behavior should be seen for 4.8 < pH < 7.5, where both polymers are appreciably charged. Because of the presence of opposite charges, the brushes may collapse onto one another or form some other kind of compact conformation at the interface.

We prepared surface-tethered polyampholyte PDMAEMA-b-PAA assemblies with one block (PDMAEMA) having a variable length and the other block (PAA) with constant thickness. We utilized the aforementioned draining method (cf. Fig. 29) in conjunction with ATRP to prepare PDMAEMA brushes with a variable length on silica substrates [109, 110]. After forming the first block, the PDMAEMA-covered substrate was immersed into an ATRP solution of *tert*-butyl acrylate (tBA); the PDMAEMA brush acted as a macroinitiator for the polymerization of PtBA with a constant length. In Fig. 31 we plot the dry thicknesses of the PDMAEMA (■) and PDMAEMA-b-PtBA (▲) as a function of the position on the silica substrate. As apparent from the data, the length of PDMAEMA varies gradually across the substrate, while the PtBA layer has approximately constant thickness across the sample, as expected. After the polymerization, the PtBA block was converted into PAA using the method described earlier in Sect. 4. Unlike the case of the PAA brush described earlier in the text, where some of the polymers may have been cleaved off the surface because of the hydrolysis of the ester group in the BMPUS initiator, no cleavage is expected to occur here because the PDMAEMA block is fairly thick. This is further justified by exploring the data in Fig. 31, where the dry thickness of PAA is about 40% of that of PtBA (compare the PDMAEMA-*b*-PAA thickness (●) with the PDMAEMA-*b*-P*t*BA thickness (▲)); the expected thickness reduction is about 52% based on the volume change associated with cleaving the bulky *tert*-butyl group from each *t*BA monomer [111].

The response of the PDMAEMA-*b*-PAA brushes to changes in pH was elucidated from measurements performed with spectroscopic ellipsometry (SE). The SE experiments were conducted in a similar fashion as described earlier in the text. Specifically, the samples were immersed into a solution of a given pH and a constant ionic strength (≈ 0.001 M) and allowed to incubate in an ellipsometric cell for at least 5 h before each measurement. The measurements were carried out on several fixed positions along the sample; these are marked by numbers 1 through 4 in Fig. 31. In Fig. 32 we plot the cosine of Δ and tangent of Ψ as a function of the wavelength of incident polarized light, λ, at two positions on the specimen: position 1 (distance = 7.5 mm on the substrate) and position 4 (distance = 21.0 mm on the substrate). At each position, the data has been collected at various values of pH ranging from 3.52 to 9.50. A visual examination of the raw data indicates a clear effect of pH on the system behavior. For example, take the data collected at position 1 and follow the behavior of the peak in cos(Δ). As pH increases from 3.52, the position of the peak starts to move towards higher wavelengths. For pH > 4.84, the behavior reverses and the peak position moves towards smaller wavelengths. Finally, at pH > 7.85, the peak starts to shift towards higher wavelengths again.

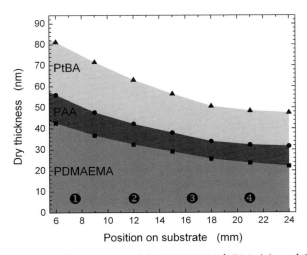

Fig. 31 Dry thicknesses of PDMAEMA (■), PDMAEMA-*b*-PAA (●), and PDMAEMA-*b*-P*t*BA (▲) as a function of the position on the silica substrate. The symbols ①, ②, ③, and ④ mark the loci on the specimen where the wet thickness measurement was carried out (cf. Figs. 32–34)

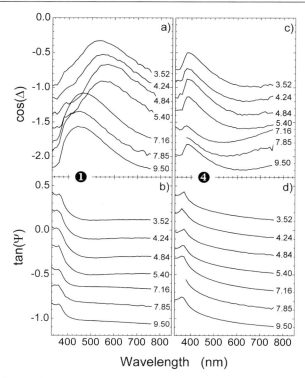

Fig. 32 Dependence of cosine of Δ and tangent of Ψ, where Δ and Ψ are ellipsometric angles related to the change of amplitude and phase shift of the incident polarized light, on the wavelength of the incident polarized light collected from PDMAEMA-*b*-PAA brushes immersed in solutions of a constant ionic strength (≈ 0.001 M) and pH ranging from 3.52 to 9.50. The data in figures a and b (c and d) have been collected at the position 1 (4). For clarity the data for $\cos(\Delta)$ and $\tan(\Psi)$ collected at pH > 3.53 were shifted vertically by – 0.2 relative to each previous set

One of the drawbacks of ellipsometry is that the raw data cannot be directly converted from the reciprocal space into the direct space. Rather, in order to obtain an accurate ellipsometric thickness measurement, one needs to guess a reasonable dielectric constant profile inside the sample, calculate Δ and Ψ and compare them to the experimentally measured Δ and Ψ values (note that the dielectric profile is related to the index of refraction profile, which in turn bears information about the concentration of the present species). This procedure is repeated until satisfactory agreement between the modeled and the experimental values is found. However, this trial-and-error process is complicated by an ambiguity in determining the "true" dielectric constant profiles that mimic the experimentally measured values. In what follows we will analyze the data qualitatively and point out trends that can be observed from the experimental measurements. We will demonstrate that this process still offers a qualitatively correct picture of the conformational response of the PDMAEMA-*b*-PAA brush assemblies to changes in pH. A more quantitative modeling will be presented in a separate publication (Tomlinson et al., 2005, unpublished results).

To proceed, we pick a wavelength λ and plot the values of Δ and Ψ as a function of pH at positions 1 through 4 on the specimen. In Fig. 33 we present such a plot for $\lambda = 630$ nm, which is close to the wavelength of red light (= 632.8 nm). The trends in Δ and Ψ are very similar for positions 1–3 and differ for position 4. We note that similar trends in Δ and Ψ are detected at wavelengths $\lambda > 600$ nm, i.e., the spectral region to the right of maximum in Δ. We concentrate on the trends seen at one of the first three positions, say position 1, and describe the behavior in Δ and Ψ. While Δ is very sensitive to the structure of ultrathin layers [112], values close to $\Delta = 180°$ may be flawed by a large error due to reading in the analyzer [113]. From the data in Fig. 33, one can see that for a given position Δ (or Ψ) decreases as pH increases, reaches a minimum at pH = 4.84, then starts to increase, peaks at pH $\approx$ 7.85 and then levels off. The behavior at the positions 2 and 3 is similar to that at the position 1. In order to relate the changes in Δ (or Ψ)

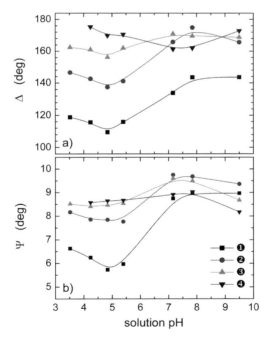

Fig. 33 Dependence of Δ and Ψ on the solution pH for PDMAEMA-*b*-PAA brushes determined from the curves shown in Fig. 30 at the wavelength of 630 nm at various positions along the specimen, x: (■, x = 7.5 mm ①), (●, x = 12 mm ②), (▲, x = 16.5 mm ③), and (▼, x = 21 mm ④). The *lines* are meant to guide the eye

to the variation in the real thickness of the brush, pick a value of pH (say pH = 3.51) and concentrate on the values of Δ (or Ψ) as a function of the position on the sample. Recall, that the total thickness of the brush is the highest at the initial position along the specimen and decreases as one traverses across the specimen towards higher values of x (cf. Fig. 31). Note that while in the data presented in Fig. 31 we plot the dry thicknesses, the values of Δ and Ψ shown in Fig. 33 correspond to wet thicknesses, which are, on average, about 6–10 times larger for fully swollen chains [66, 114–116]. Since at pH = 3.52 PDMAEMA is charged and PAA is expected to be nearly neutral, the thickness trends seen in Fig. 33 will be magnified. The degree of polymerization, and hence the degree of stretching, will be the highest at position 1 and will decrease as one traverses the sample towards positions 2, 3, and 4. From the data in Fig. 33, one can see that the decrease in polymer thickness is accompanied with an increase in Δ (or Ψ). Hence for this combination of the dielectric constants and the range of brush thicknesses investigated there, the increase in Δ (or Ψ) corresponds to the decrease in the overall brush thickness. The information about the trends in Δ (or Ψ) can be used to interpret the total brush thickness in the sample and to generate a tentative picture of the brush behavior as a function of the solution pH.

   The cartoon in Fig. 34 is a schematic representation of the brush conformations at positions 1–3. At low pH, the PDMAEMA block is fully charged while there are only a few charges present in the PAA block. Due to the repulsions among the positive charges along the PDMAEMA block, the block copolymer extends far into the solution. With increasing pH, the number of charges along the PDMAEMA block decreases, albeit only slightly because the $pK_{1/2}$ of PDMAEMA is still very much afar [89, 108]. Simultaneously, there is

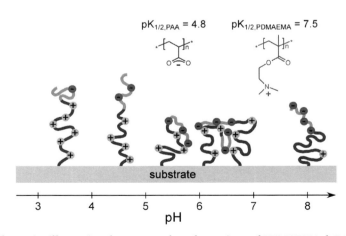

Fig. 34 Schematics illustrating the suggested conformations of PDMAEMA-b-PAA brushes of a constant composition grafted on a solid substrate as a function of the solution pH (cf. Figs. 31 and 33)

an increase in the number of negative charges generated along the PAA block. The presence of both charges should cause the brush to stretch even further away from the surface. However, looking back at the data presented in Fig. 32, we notice an increase in Δ for $4.8 < pH < 7.8$. This behavior of Δ corresponds to a decrease in the thickness of the brush. We hypothesize that within this pH window, the polymers may form some kind of compact/collapsed structure [101], with individual chains self-collapsing or folding down on their neighbors, as schematically depicted in Fig. 34. Finally, at $pH > 7.8$, the number of charges along the PDMAEMA block is dramatically reduced, while the upper PAA block becomes almost completely charged, which leads to a slight increase in the brush length. As stated earlier, the behavior at position 4 differs markedly from that observed at positions 1 through 3. One has to be careful about making conclusions about the chain conformation from Δ alone because, as mentioned earlier, values of Δ close to 180° are associated with an appreciable uncertainty. Thus in this case, we will explore only the Ψ values (cf. Fig. 33b). We note that the trends in Ψ as a function of pH are similar to those observed at positions 1–3, indicating that the proposed conformational changes depicted in Fig. 34 are still valid. Some small variation may occur, however. The brush collapse seen at $4.8 < pH < 7.8$ for polymers at positions 1–3 may not be as pronounced. In addition, the rapid decrease in Ψ for $pH > 7.8$ indicates a stronger stretching of the brush. Theses two trends may be tentatively explained by the fact that while the number of charges along PAA remains relatively constant, the density of charges on PDMAEMA decreases because of the decreasing molecular weight of the bottom block. There may thus be an imbalance in opposite charges, with the number of negative charges dominating. As a result, the PDMAEMA-b-PAA brushes may not fully collapse for $4.8 < pH < 7.8$ and will extend deeper into the solution for $pH > 7.8$.

The example presented and discussed in this section of the paper illustrates the complexity of the interfacial behavior of surface-tethered polyampholyte brushes. The presence of two opposite of charges and the variability in their number gives rise to a series of various complex conformations of polyampholytic diblock copolymer brushes. In order to shed more light on the performance of surface-grafted polyampholyte assemblies, one has to systematically vary the individual system variables, i.e., the number of charges and the degrees of polymerization of each block, and also the brush grafting density. This is where the combinatorial design may become particularly beneficial. In the case we just discussed, the combinatorial approach proved helpful as it offered a convenient means of qualitatively interpreting the ellipsometric data. In addition, the present example also reveals that in order to understand the polymer behavior in such a complex system, multiple characterization tools that provide complementary information should be applied. For example, the ellipsometry measurements can be complemented with experiments that are capable of "visualizing" the polymer conformations at solid/liquid interfaces, such as scanning force microscopy, or measurements using surface forces apparatus.

8
Orthogonal Grafted Polymer Gradients

Orthogonal gradient substrates are those in which two material properties vary continuously along two mutually perpendicular substrate dimensions. Since these two properties are changing continuously in two orthogonal directions, every possible combination of the two properties can be probed by using such a set-up. Working with orthogonal gradient substrates not only saves time and resources but it also minimizes the systematic error associated with doing individual experiments. We have established in the previous sections that one can independently control the grafting density (σ) and molecular weight (MW) of grafted polymer brushes. These methods of

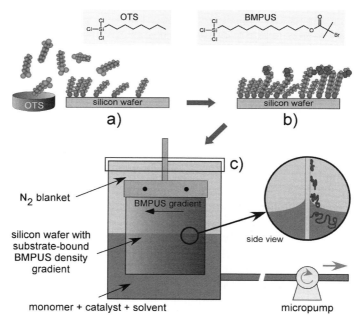

Fig. 35 Schematic illustrating the formation of an orthogonal poly(2-hydroxyethyl methacrylate) (PHEMA) gradient. A molecular gradient of n-octyltrichlosilane (OTS) is formed on a silica-covered surface (**a**), and the empty spaces on the surface are filled with [11-(2-bromo-2-methylpropionyloxy) undecyl] trichlorosilane (BMPUS). The substrate, BMPUS gradient positioned horizontally, is placed in the custom-designed polymerization chamber and PHEMA molecular weight gradient is generated using the draining method, as described in the text. (Reproduced with permission from [164])

forming separate gradients can be combined to form orthogonal polymer brushes in which σ and MW vary continuously along two perpendicular directions [117, 118].

Fabrication of a surface-grafted σ-MW orthogonal gradient entails two steps: (1) formation of a concentration gradient of the initiator molecules; and (2) growth of surface-anchored chains with a molecular weight gradient in a direction perpendicular to that of the initiator concentration gradient (cf. Fig. 35). To accomplish the first step we use the organosilane vapor diffusion technique described in Sect. 3. A gradient of initiator BMPUS molecules was formed by the backfilling method described in Sect. 4. Growing chains from such an initiator concentration gradient results in a grafting density gradient of grown brushes along the direction of the initiator gradient (X-direction) [76]. To achieve MW gradient in a direction perpendicular to that of σ, we rotated the sample by 90° and immersed it in a polymerization mixture (along the Y-direction) placed in a de-aerated reaction chamber. By continuously removing the polymerization medium from the chamber a MW gradient on the substrate was generated in the Y-direction [94]. The MW of grafted polymer chains at various points along the X-direction for a given Y should be approximately the same since all those points remain in the reaction media for the same period of time [118]. We have succeeded in creating an orthogonal gradient of grafted PDMAEMA using the methodology described above. In Fig. 36 we plot the dry thickness profile of grown PDMAEMA in X (σ gradient) and Y (MW gradient) directions. As mentioned earlier, the dry thickness of a grafted polymer is given by $h = \sigma M_n/(\rho N_A)$;

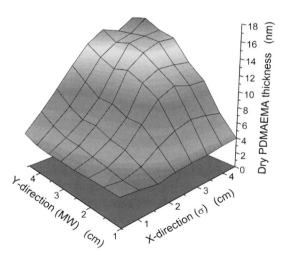

Fig. 36 Dry thickness of poly(dimethyl aminoethyl methacrylate) (PDMAEMA) on an orthogonal PDMAEMA gradient grafted on a silica substrate. Thickness was measured in a grid of 5 mm × 5 mm on the substrate. (Reproduced with permission from [117])

thus, h depends on both MW and σ of the grown polymer. For a given value of Y on the sample, h increases continuously as one moves away from the OTS end along the X direction. A similar trend in thickness is observed along the Y direction for a given X as one moves from a region that was in the reaction media for a shorter period to a region that stayed in the mixture for a longer time. These measurements indicate that an orthogonal gradient was successfully formed.

Orthogonal gradients, when combined with a large pool of functional polymers, offer powerful combinatorial platforms for studying complex phenomena; they can be tailored to serve as separation devices or as templates for organization of nonpolymeric materials. In the following section, we will elaborate further on some of these applications.

9
Applications of Surface-Grafted Polymer Gradients

In the previous sections, we showed the utility of polymer gradients in enhancing our understanding of (1) the behavior of polymers in confined geometries, especially their response to ionic solutions and (2) kinetics of growth of grafted polymers. Apart from these applications, the grafted gradients can be used as templates for distributing nonpolymeric objects. Specifically, we will show that end-anchored polymer gradients can be exploited to gain insight into the physics underlying the formation of brush-nanoparticle composite assemblies and adsorption of proteins on brushes.

9.1
Nanoparticle Assemblies on Gradient Polymer Brushes

Driven by the promised goals of nanotechnology, the last decade saw an outburst in the manipulation and control of materials on a "small" scale [119, 120]. This high volume of activity has been fuelled, in part, by the potential for exploiting unique properties of nanostructures in commercially important applications, such as sensors [121], high efficiency solar cells [122], single molecule detectors [123] and electroluminescent devices [124]. Many of these functional devices utilize nanoparticles, whose typical dimensions are intermediate between mesoscopic and molecular systems. Consequently, they offer a wide range of unusual properties not found in either mesoscopic or molecular systems [125]. Fabrication of structures comprising organized arrays of nanoparticles embedded in material matrices represents one of the most important challenges facing today's materials scientists and engineers. The macroscopic properties of nanoparticle-based composites will reflect both the characteristics that are specific to nano-objects [126, 127] (e.g., size-dependent optical and electronic properties, ability to form ferromagnetic monodomains, optical microcavities or to generate third-order harmonic optical waves) as well as those that are characteristic of the large-mesh periodic structure of nanoparticles in the matrix (e.g., a possible coherent response to electromagnetic radiation) [128–130]. In order to utilize nanoparticles as components of functional devices, it is vital to develop methods for placing particles into chemically and structurally well-defined environments. This is typically achieved through covalent or electrostatic interaction of particles with a substrate-bound molecular or polymer film [131–134]. For example, Natan and coworkers immobilized gold nanoparticles on self-assembled monolayers (SAMs) of amino- or mercaptosilanes bound to a silica substrate. Adsorption of particles by this method resulted in only a sub-monolayer coverage, further adsorption on the surface being limited by repulsion between the charged particles. Often, such low surface coverage does not produce the property enhancement expected due to the presence of the nanoparticles, thus limiting the practical utility of the nanoparticle-based devices.

One method to increase surface coverage and loading of particles on the substrate is to generate three-dimensional assemblies of particles using an external matrix such as a polymer. Due to the several advantages that polymers have over the classical metal-, ceramic- or semiconductor-based matrices, 3-D composites based on nanoparticle/polymer dispersions may constitute the next generation structures for the aforementioned applications. Polymers are usually optically transparent, possess insulating properties, and are inexpensive and easy to process. Additionally, they can stay in a solid glassy state or behave as viscous fluids, depending on whether temperature is below or above their glass transition temperature. This solid/fluid duality of polymers allows one to control the mobility of guest nanoparticles embedded in polymeric matrices by simply adjusting the processing temperature or the content of a plasticizer. Typical examples of using polymers to form surface-bound nanoparticle assemblies include the use of block copolymer matrices [135, 136], or polymers tethered chemically to the substrate [137]. Polymer brushes offer model environments for studying the organization of nanoparticles within polymer matrices. A simple homopolymer as well as copolymer brushes can be made with the desired molecular weight and grafting density; they allow easy manipulation of polymer conformation by either changing solvent quality or temperature; they facilitate the effect of the interface to be studied. Among the many parameters that govern dispersion of the nanoparticles inside the polymer brushes are: (1) particle size; (2) strength of polymer/particle interaction [137]; (3) polymer grafting density [138]; (4) polymer chain length [139]; and (5) solvent quality. The large number of parameters makes probing the behavior of brush/nanoparticle hybrids very difficult. The various polymer gradients described earlier in the report can be meticulously applied to fully understand the intricacies of brush/particle composites (cf. Fig. 37). In what follows, we will describe our progress towards this goal.

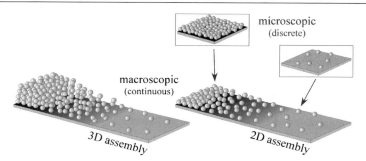

Fig. 37 Schematic illustrating the organization of nanoparticles in two- (2-D) and three-(3-D) dimensional gradients and the dual character of such assemblies – continuous on mesoscale and discontinuous on micro- and nanoscales. (Reproduced with permission from [140])

9.1.1
Nanoparticles on Polymer Brushes with a Molecular Weight Gradient

In this section, we investigate the effect of polymer molecular weight on nanoparticle attachment to the brushes, while keeping all other polymer and particle parameters constant [140]. For this purpose, we grew PAAm brushes with a gradient in molecular weight using the methodology described in Sect. 5. PAAm was chosen because its backbone has abundant amino groups, which, when ionized, are known to have affinity for negatively charged gold nanoparticles used in this work [132, 134, 141]. Citrate-capped gold nanoparticles with an average particle diameter of 16.9 ± 1.8 nm were synthesized following the recipe by Fren [142]. The prepared gold nanoparticle solution is slightly acidic in nature (pH ≈ 6.5) and the particles are negatively charged due to the presence of an anionic protective coating on the particles. The dry thickness profile of grown PAAm brush along the substrate as measured by SE is shown in the bottom panel of Fig. 38. PAAm thickness along the substrate varies initially almost linearly and then it slowly levels off. The latter behavior, consistent with literature reports of nonlinear growth of PAAm brushes, can be attributed to the premature chain termination [143]. Since the polymerization initiators are distributed homogeneously on the substrate, the grafting density of the polymerized PAAm chains can be presumed to be roughly equal across the substrate. Accordingly, a gradient in measured dry thickness of PAAm is attributed primarily to a gradient in chain length (or equivalently molecular weight) of the end-grafted PAAm.

The PAAm chain length gradient so prepared is kept immersed in a colloidal gold solution overnight to achieve complete immobilization of particles in the brush. The specimen is subsequently removed from the gold solution, washed thoroughly with DI water and blow-dried with nitrogen. On the basis of our experience with attachment of gold nanoparticles to the amino group-

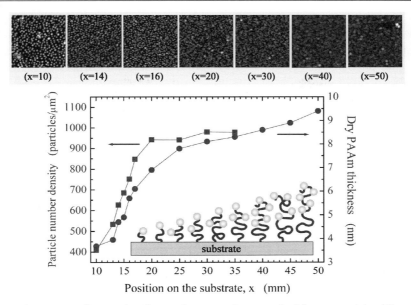

Fig. 38 (*Upper panel*) Scanning force microscopy images of gold nanoparticles (diameter ≈ 17 nm) adsorbed along a surface-anchored poly(acryl amide) brush with a molecular weight gradient (Edge of each image = 1 μm). (*Lower panel*) Dry thickness of poly(acryl amide) on the substrate before particle attachment (*right*, ●) and particle number density profile (*left*, ■). (Reproduced with permission from [140])

containing SAM [134], we speculate that particles are held tightly to PAAm brushes through electrostatic interactions between the charged citrate (on the particle) and $- NH_3^+$ (on the PAAm chain) groups. Specifically, while at low pH nanoparticles are attached to the polymer brushes, when deposited from a gold solution of pH > 9, almost no gold nanoparticles are detected inside the polymer brush. It should be noted that in addition to the electrostatic interaction, other forces such as hydrogen bonding interaction between particles and the charged polymer might also be partly responsible for driving particle attachment to the brush.

When the PAAm brush is immersed in an aqueous colloidal gold solution, swelling of the brush occurs, as water is a good solvent for PAAm. The brushes may swell up to as much as 6–10 times their dry thickness [66, 114–116]. This swelling introduces sufficient space in the structure for the particles to explore deep into the brushes. The actual depth to which a particle can delve into the brush depends on the interplay between the chain length, grafting density of the brush, particle size, particle/polymer interactions, and the solvent quality [144]. Generally, swelling polymer brushes beyond their equilibrium conformations is entropically prohibited. This entropic penalty will increase with increasing the polymer grafting density, brush length, and the size of the particle [144]. However, this entropic barrier can be overcome if strong interactions between the polymer and particles exist, such as electrostatic interaction between the negatively charged ions on the gold particle surface and $-NH_3^+$ groups along the PAAm backbone. In the PAAm/gold/water system, the number of sites available for the gold particle attachment is proportional to the number of accessible charges along the PAAm backbone. Thus, one may expect that the number of attached nanoparticles will increase with increasing PAAm brush molecular weight. When the substrate is removed from the aqueous solution and dried, the swollen brush collapses, thus bringing down particles along with it, as shown in the schematic in Fig. 39. Such structural collapse makes the dry structure very different from the wet one. While understanding of the wet structure requires in-situ characterization, it is relatively easier to characterize dry samples with a battery of ex-situ thin film characterization techniques. The main objectives of film characterizations were to establish the concentration of particles as a function of PAAm molecular weight and their distribution in the PAAm dry brush as a function of the particle size.

SFM was used to determine the particle number density along the substrate. Topographical images from selected regions on the substrate are shown in the top panel of Fig. 38. The concentration of the particles, as measured by SFM, increases with increasing brush thickness. In the bottom panel of Fig. 38, we plot the dry thickness of the PAAm brush before the particle attachment (right ordinate) and the number of particles attached per unit area (left ordinate) plotted as a function of the position on the substrate. Proportional increase in both quantities along the sample lends credence to our assertion that the gradient in particle density is due to the gradient in the

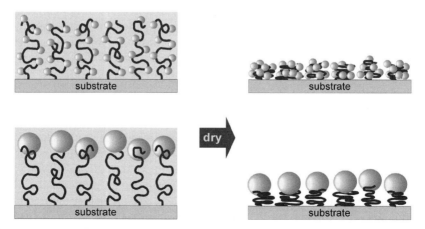

Fig. 39 Schematic illustrating the structure of a wet (*left*) and dry (*right*) brushes with attached small (*top*) and large (*bottom*) particles. (Reproduced with permission from [140])

anchored PAAm molecular weight. As can be seen from the first few SFM images on the left-hand side of the top panel, particles are well separated and can be counted individually for low brush thickness. In contrast, for particles deposited on thicker brushes the images become fuzzy and the particle densities cannot be evaluated easily. Experiments performed with several different SFM tips confirmed that this observation is not an artifact due to tip damage during scanning. Hence, the results from the SFM experiments lead to the speculation that while at low brush thickness particles predominantly adsorb on the surface of the brush, for thicker brushes the "fuzziness" is due to the onset of nanoparticle penetration into the brush.

Further independent evidence confirming the gradient nature of the substrate comes from UV-Visible (UV-Vis) light absorption spectroscopy performed on a particle gradient [140]. For this set of experiments, the procedure of forming PAAm brushes with a chain length gradient and subsequent particle attachment as detailed earlier is repeated for a glass substrate. Our results (not shown here) reveal that as particle concentration increases along the gradient, i.e., as the x coordinate along the gradient increases, the intensity of the plasmon absorption peak (around 530 nm) associated with the gold nanoparticles [126] also increases, accompanied by a red shift in the peak position. Traditionally, such a behavior has been associated with an increase in gold particle concentration [131, 145, 146], thus justifying our claim that we form a particle density gradient by using the polymer brush molecular weight gradient. Increased broadening of the plasmon peak and increased tail absorption in the red region of the spectrum on thicker PAAm brushes suggest that the inter-particle interaction intensifies. However, we discount any aggregation of particles in the absence of a new low-energy peak in the absorption spectrum. Thus, the particles still maintain their individual character even at higher loading on thick brushes.

One outstanding issue in polymer brush/nanoparticle composite systems involves determining the spatial distribution of particles in the brush. There have been only a few studies concerning the use of polymer brushes to control the morphological behavior of nanoparticles. Levicky and coworkers formed a polymer brush by spin coating and annealing poly(styrene-b-ethylene propylene) block copolymer on top of polystyrene homopolymer [139]. The poly(ethylene propylene) block, which formed the brush, was incompatible with octanethiol-coated gold nanoparticles, thus causing particle aggregation on the surface of the brush and consequently rendering particles incapable of penetrating into the brush. Cohen Stuart et al. used poly(styrene-b-ethylene oxide) block copolymers, whose poly(ethylene oxide) blocks exhibited favorable interaction with silica particles. The researchers carried out a systematic study of particle intake in the brush as a function of the grafting density and length of the brush [137]. However, the issue of the spatial arrangement of particles in the brush was not addressed in [137]. We have recently reported on a series of angle-dependent X-ray photoelectron spectroscopy

(XPS) experiments in order to determine the concentration of gold colloids inside PAAm brushes [140]. Specifically, we studied dispersions of two different sizes of gold particles (diameters ≈ 3.5 nm and ≈ 16 nm) in homogenous PAAm brushes. Smaller gold nanoparticles were prepared by following the recipe in [147]. In Fig. 40 we plot the gold-to-carbon (Au/C) elemental ratio as a function of the XPS take-off angle for the two nanoparticle sizes in PAAm brushes with two different thicknesses. The data reveal that for both particle sizes the Au/C ratio in thin (≈ 5.5 nm) PAAm brushes decreases as the take-off angle increases. Considering that the increase in the take-off angle is associated with increasing the probing depth in the sample, these results indicate that gold nanoparticles are located predominantly closer to the PAAm brush/air interface. The Au/C ratio for the larger gold particles in thicker (≈ 13 nm) brushes also decreases with increasing the take-off angle, but relative to the thin PAAm case the absolute Au/C values are larger and the difference between the smallest and largest take-off angle decreases. This behavior suggests that more particles are incorporated into the brush and that the particles may start to form a 3-D assembly. The XPS results collected from the 3.5 nm gold particles in thicker (≈ 10 nm) PAAm show a trend in the Au/C ratio opposite to that of the larger particles. Specifically, upon increasing the take-off angle, the Au/C ratio is found to increase, suggesting that the particles form a 3-D assembly by dispersing deeper into the PAAm brush.

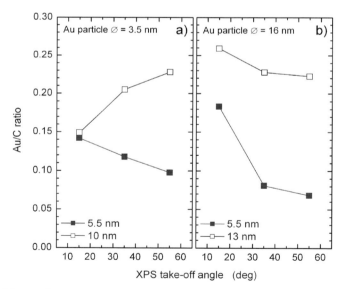

Fig. 40 Gold-to-carbon ratio as a function of the XPS take-off angle measured on gold nanoparticle assemblies on PAAm for nanoparticles having a diameter of 3.5 nm (**a**) and 16 nm (**b**). The dry thickness of bare PAAm is given in the legend. The *lines* are meant to guide the eye. (Reproduced with permission from [140])

These results illustrate the intimate interplay between the particle size and the brush thickness that determines the spatial distribution of the particles inside the polymer brushes.

Comparing our experimental results with the theoretical predictions of particle dispersion in polymer brushes provides more insight into the behavior of nanoparticles in thin polymer films. Kim and O'Shaughnessy predicted [144] that for particles having favorable interactions with the polymer brush, the extent of dispersion of particles within the brush depended on particle size. For a given N and σ of the brush, they identified three regimes of particle penetration in the brush. In the first regime, which the authors identified as the "mixing regime", very small particles were predicted to disperse freely within the polymer film. Above a threshold particle size b^*, equilibrium particle penetration was limited to a depth inversely proportional to the particle size; the smaller the particle, the greater was the penetration in the brush. This second regime was called the "partial mixing" regime. The third regime, called the "exclusion regime", considered the case when the particle size was greater than a second threshold b_{max} such that the particles just stayed on the surface of the brush. As expected, b^* and b_{max} depended on N and σ. For a given particle size and brush grafting density, particles were found to shift from the "exclusion" regime to the "partially mixed" regime upon increasing the brush chain length. Our angle-dependent XPS experiments of gold nanoparticle dispersion in PAAm brushes conform to the theoretically predicted trends.

9.1.2
Orthogonal Nanoparticle Gradients

In the previous section, we showed how one directional grafted polymer gradient was used to study the effect of polymer molecular weight on attachment of nanoparticles to the brush. In this section, we will demonstrate how orthogonal gradients of anchored polymers can be utilized to study the effect of molecular weight as well as grafting density simultaneously *on one single substrate*. For this purpose, we prepared an orthogonal gradient of surface-anchored PDMAEMA brushes using the procedure described in Sect. 8. The study of attachment of citrate-covered gold particles on an orthogonal PDMAEMA gradient was carried out with two objectives: (1) to study how *MW* and σ affect the loading of particles in the brushes; and (2) to visualize the orthogonal polymer gradient. This visualization is made possible by the strong visible light absorbance associated with surface plasmon resonance of gold nanoparticles [126, 131].

In Fig. 41 we show a photograph of the orthogonal gradient upon particle uptake in the brush. The color of the slide changes gradually from light pink (region of low *MW* and low σ) to dark violet blue (region of high *MW* and high σ) as one moves along the directions of increasing *MW* and σ. This color

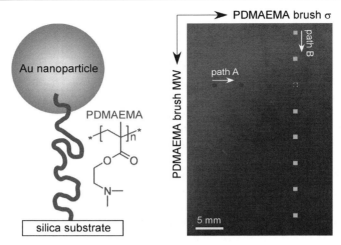

Fig. 41 (*left*) Schematic illustrating the attachment of citrate-covered gold nanoparticles to a poly(dimethyl aminoethyl methacrylate) (PDMAEMA) polymer brush grafted on a silica substrate. (*right*) Photograph of a glass slide showing gold nanoparticles (≈ 17 nm in diameter) bound to an orthogonal gradient of surface grafted PDMAEMA. Note the color variation along the direction of molecular weight (*MW*) as well as grafting density (σ) gradients, clearly indicating different gold particle uptake along each gradient. (Reproduced with permission from [117])

variation is indicative of inter-particle plasmon coupling associated with an increase in uptake of particles in the brush as *MW* (or σ) increases [126, 131]. In Fig. 42, we plot visible light absorbance spectra collected along σ (path A) and *MW* (path B) gradients. There are two points worth noting in these plots: (1) increase in intensity of the plasmon absorbance peak in the direction of increasing *MW* or σ; and (2) the concomitant red shift of the plasmon peak position. The first feature is due to the increasing number of particles attached to the polymer chains, whereas the second feature suggests intensified inter-particle plasmon coupling accompanying the nanoparticle crowding on the substrate [126]. Such a correlation between plasmon peak intensity and the number of adsorbed particles has been well documented in the literature [131, 140].

It is instructive to compare the experimental work with theory describing the favorable interaction of particles with brushes. Currie et al. studied the formation of a complex between grafted polymer chains and mesoscopic particles using an analytical self-consistent-field theory [18, 137, 148, 149]. They predicted that grafting longer chains enhances adsorption of the particles. Longer chains offer more adsorption sites resulting in favorable polymer-particle interaction, which outweighs the increase in osmotic pressure in the brush upon particle attachment. The number of adsorbed particles also increases as grafting density is increased in the low grafting density regime.

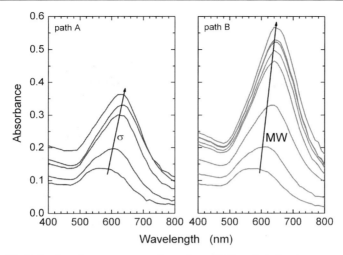

Fig. 42 Visible light absorbance spectra taken along (*left*) red circles (constant molecular MW, σ gradient) and (*right*) green squares (constant σ, gradient MW) shown in Fig. 41. (Reproduced with permission from [117])

This is because of the increased attachment sites that become available as σ is increased. However, there is a maximum in the adsorption as a function of σ. At very high σ, the osmotic pressure caused by insertion of particles in the grafted layer overcomes the favorable polymer-particle binding energy. This results in expulsion of particles from the proximal (i.e., closer to the substrate) to the distal (i.e., away from the substrate) region of the brush [137, 138]. The maximum in particle adsorption is a manifestation of the balance between the increasing number of adsorption sites and increasing osmotic repulsion as σ is increased. Our gold nanoparticle adsorption results along the MW gradient arm of the orthogonal gradient substrate are in line with the theory. However, we do not observe a maximum in the number of adsorbed particles as a function of σ. Instead, adsorption increases monotonically as σ is increased. We suspect that this is due to the large size of particles (diameter $\approx$ 17 nm) used in our study compared to the average distance between the grafted polymer chains. From our previous study reporting on the penetration of particles of different sizes into surface-anchored polymer brushes, we know that 17 nm particles do not penetrate the brush (with thickness similar to the ones considered in this study) to a great extent [140]. Larger particles tend to reside in the distal region of the brush whereas smaller particles (diameter $\approx$ 3.5 nm) are able to explore deep into the brushes and arrange themselves in a 3-D-like structure. Consequently, upon attachment to the brush, smaller particles have the potential to cause greater osmotic penalty than the larger particles considered here. Thus, for larger particles we expect a monotonous increase in particle number density as a function of σ. Our results support this assertion.

9.2
Protein Adsorption on Grafted Polymer Gradients

Polymer coated substrates are used to prevent protein adsorption and thus minimize subsequent biofouling [7, 150]. Polymer chains chemically grafted to the surface have been predicted to be more efficient at preventing protein adsorption than physisorbed polymers. Theoretical models revealed the roles of different molecular features of the grafted polymer in controlling protein adsorption [151, 152]. By combining modeling and experimental results, Szleifer and coworkers demonstrated that the main molecular parameter that governs protein adsorption is the polymer surface coverage [150, 153–155]. Others have argued that for inhibiting protein adsorption, a high grafting density is most effective for small proteins, whereas increased layer thickness is most effective for large proteins [8, 156]. While some experimental evidence of these predictions has already appeared [7, 157–159], our understanding of protein adsorption on polymeric substrates is still far from complete. We envisage that the substrates comprising grafted polymer gradients represent tools that will enable *systematic* study of the role of grafting density and polymer length on protein adsorption and separation. In this section, we present results of our preliminary experiments aimed at achieving this goal.

Hydrophilic polymers such as poly(ethylene oxide) (PEO) and PHEMA have been traditionally used to prevent adsorption of proteins on synthetic biomaterials [7, 160, 161]. We have chosen to use PHEMA in our study due to its ease of synthesis on surfaces at high grafting density by ATRP. Although not as efficient as PEO in minimizing protein adsorption, PHEMA has adequate protein resistance to serve as a model polymer to show the utility of gradient substrates in systematically studying a complex phenomenon such as biofouling. For example PHEMA was found to be useful in prevention of protein deposits on contact lenses [161, 162], and in minimizing cell adhesion in routine cell culturing [163].

In Fig. 43 we plot the dry thickness profile of a PHEMA MW gradient. MW and consequently thickness increases from the left end of the substrate to the right. The top panel shows fluorescent microscopy images taken at various positions along the PHEMA gradient substrate after fibrinogen adsorption [164]. A clear reduction in the amount of adsorbed fibrinogen is evident as the molecular weight of the grafted PHEMA is increased along the gradient. This observation is in line with theoretical prediction [8] that adsorption of larger proteins can be suppressed more efficiently by increasing the thickness of the grafted polymer film. To simultaneously study the effect of MW and σ on the adsorption of proteins, we created an orthogonal

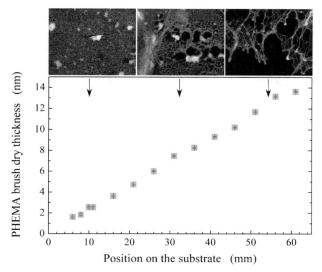

Fig. 43 Dry thickness of poly(2-hydroxyethyl methacrylate) (PHEMA) as a function of the position on the substrate (*lower panel*). The fluorescence microscopy images show the corresponding structure of fibrinogen at three positions along the PHEMA gradient (*upper panel*). (Reproduced with permission from [164])

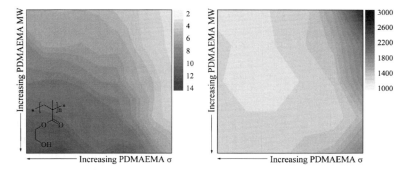

Fig. 44 (*left*) Dry thickness (in nm) of a poly(2-hydroxyethyl methacrylate) (PHEMA) brush in an orthogonal gradient as a function of the PHEMA grafting density and molecular weight. The scale represents the thickness of dry polymer (in nm). (*right*) Adsorbed amount of lysozyme as a function of the position on the orthogonal PHEMA gradient. The scale represents the fluorescence intensity (in a.u.). (Reproduced with permission from [164])

gradient in MW and σ of PHEMA [164, 165]. The thickness profile of such an orthogonal gradient is shown in the left panel of Fig. 44. In accordance with the equation $h = \sigma M_n/(\rho N_A)$, the dry PHEMA thickness increases in the directions of increasing MW as well as increasing σ, thus indicating successful formation of an orthogonal PHEMA gradient. The right panel of Fig. 44

shows a contour plot of fluorescent counts measured at various positions on the orthogonal gradient surfaces after lysozyme attachment. The intensity of fluorescence coming from the protein decreases rapidly as both MW and σ are increased, thus clearly indicating the decrease in protein adsorption. Recently, we employed PHEMA linear and orthogonal substrates to tailor the adhesion of osteoblastic cells on substrates [165]. While still at the preliminary stage of development, these experiments have confirmed the feasibility of utilizing polymer gradients as a simple means of studying biological phenomena such as protein adsorption and cell adhesion. More systematic work aimed at quantitatively characterizing protein adsorption and cell adhesion on polymer gradient substrates is currently underway.

10
Outlook

In this review we described several methods of preparing polymer brush assemblies with a gradual variation of selected properties, such as the grafting density, molecular weight, and chemical composition. We discussed how these structures could be utilized to advance our understanding of the basic physico-chemical phenomena occurring in such macromolecular systems. We also outlined several applications where the use of a gradient anchored macromolecular system is desired. We conclude this paper by outlining three avenues, in which macromolecular assemblies with property gradients may play a pivotal role in the near future.

10.1
Combinatorial Studies

One of the major application areas of gradient polymer assemblies involves high-throughput (combinatorial) studies of the interfacial behavior of molecules and macromolecules. In order to understand the real benefits of the polymer gradient approach, consider an example involving protein adsorption on substrates covered with grafted polymers. When studying the effect of brush chain length (N) and the grafting density (σ) on protein adsorption, using, for example, five discrete measurements of each variable, one has to synthesize 25 different samples and carry out 25 individual experiments, each of which is associated with its own individual error. Moreover, one needs to interpolate in order to gain information on behavior in the property space outside these 25 discrete data points. The gradient technology removes these limitations. Instead of 25 discrete samples, one can fabricate a *single* substrate, in which N and σ vary in two orthogonal directions, and use such a specimen to study protein adsorption as a function of two individually varying properties, i.e., graft polymer density and chain length, in a *single experiment*. The result of such an experiment will be a continuous response that eliminates the need for interpolation between single discrete data points. In addition, since only one sample is used, which is prepared in a single batch, the errors associated with the results are consistent across the entire property space. This is in contrast to the discrete stochastic errors associated with the individual experiments. Hence, the gradient geometry offers a novel "combinatorial" platform, which enables complete and systematic exploration of the broad parameter space during protein adsorption while simultaneously improving the efficiency of the screening process by increasing the speed and reliability. More importantly, the multi-gradient (= orthogonal, in the case of two independent gradient structure) approach offers the possibility of studying the simultaneous effect of multiple variables on the system performance. For example, by utilizing the aforementioned orthogonal gradients comprising polymer brushes with gradual variation of grafting density and molecular weight as the test substrate in protein adsorption studies, one would not only be able to interpret the relative importance of the grafting density vs. molecular weight on protein adsorption but shed more light on cooperative effects originating from the two properties [7–9, 150, 153–156, 165].

The benefits of utilizing combinatorial methods for investigating polymer properties have been outlined recently [19, 166, 167]. Polymer gradient brush assemblies are expected to play an active role in further combinatorial material effort. Possible areas of interest include (but are not limited to): study of phase behavior (stability) in liquid [168] and polymer blend [169] systems, morphological transitions in block copolymers [170, 171], cell culturing [58, 172], and others.

10.2
Templating and Material Processing

Gradient polymer brush assemblies will play an important role in the development of novel methods of (1) controlling the interfacial chemistry of surface-anchored macromolecules and (2) material templating. This activity will have an impact on the fields of polymer physics and polymer chemistry and also newly emerging areas of nanoscience and nanotechnology. Gradient polymer brush structures will allow for the study of phenomena that have not been accessible previously or are not easy to study using conventional methodologies. For example, using the combinatorial design outlined in this paper one would be able to systematically probe the interfacial performance of surface-grafted copolymers as a function of their chemical composition or analyze the effect of charge distribution in polyampholyte brushes on conformational changes of these complex macromolecules. Methodologies involving time-dependent sample exposure to polymerization solution will provide information about the kinetics of polymerization in confined geometries. The same concept may also assist in systematically establishing the effect of the macroinitiator length and chemistry on the polymerization kinetics of living/controlled radical polymerizations. Generating polymer-based gradient assemblies will provide convenient routes for exploring the possibility of using surface-grafted polymers as soft matter scaffolds for organizing nano-sized objects, such as nanoparticles. The gradient nature of the template will assist in establishing interplay among the particle size, particle/polymer interactions, the brush height, and the brush grafting density on the spatial organization of the nanoparticles [144]. Developing procedures leading to the reproducible production of structurally well-defined nanocomposites will lead to materials with unique properties which may be used in a variety of next-generation devices, examples of which include electronic or optical devices, specialty coatings, sensors, or magnetic storage media.

10.3
Directed Transport of Host Objects

Utilizing polymer gradients will provide a simple and convenient means of studying the dynamics of various molecular and biomolecular phenomena, in particular the motion of nano-to-micro sized objects. The gradient structures are expected to provide a new paradigm for facilitating phenomena in which motion plays an important role, such as assembly or a separation. One beautiful example of such a utilization of the gradient nature of grafted polymers and their role in controlling the motion of a small object are the cylindrical molecular brushes with a grafting density gradient of the grafted side chains (cf. Fig. 45). Collaboration between the Matyjaszewski and Sheiko groups resulted in establishing the conditions leading to the controlled motion of these "synthetic caterpillars" on surfaces [173]. When properly

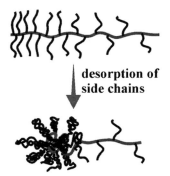

desorption of
side chains

Fig. 45 When cylindrical brushes possessing a gradient grafting density along the backbone are adsorbed on a surface, one can observe a transition from a rod-like to a tadpole conformation upon partial desorption of side chains. The end with a higher grafting density, and thus with a greater extension of the side chains, is predicted to collapse more readily than the loose end (Reproduced with permission from [173])

engineered, surface-tethered polymers can also serve as active elements in controlled surface transport of materials and thus play the role of soft-matter "engines" directing the motion of nano-objects. The continuous variation of the physico-chemical properties of the brushes provides the driving force for the motion of the target object, such as particles. Hence, employing polymer gradients will offer a simple and yet robust means of studying molecular and biomolecular phenomena, such as adsorption, and separation. Utilizing the findings in this field may, in turn, lead to designing novel tools that can be integrated rapidly into technology, such as diagnostic or property screening devices.

Acknowledgements The authors gratefully acknowledge financial support from the National Science Foundation (Grants CTS-0209403, CTS-0403535, and EEC-0403903) and The Camille & Henry Dreyfus Foundation. The NEXAFS experiments were carried out at the National Synchrotron Light Source, Brookhaven National Laboratory, which is supported by the U.S. Department of Energy. The SE was purchased with funds awarded to JG through the NSF's Instrumentation for Materials Research Program (Grant No. DMR-9975780). The work reviewed here would not have been possible without numerous collaborative efforts with several researchers. We thank Dr. Kirill Efimenko (NCSU) and Dr. Daniel Fischer (NIST) for their assistance during the course of the NEXAFS experiments, Dr. Igal Szleifer and Mr. Peng Gong (Purdue University) for the theoretical work involving surface-tethered charged systems, Dr. Andrea Liebmann-Vinson, Mr. Bryce N. Chaney, and Mr. Harry W. Sugg (Becton Dickinson Technologies) for the XPS work and experiments involving protein adsorption on gradient polymer surfaces. Dr. Petr Vlček and Dr. Vladimír Šubr (Institute of Macromolecular Chemistry, Prague, Czech Republic) are thanked for their numerous contributions in characterizing polyelectrolyte macromolecules. We also thank Professor Michael Rubinstein (University of North Carolina at Chapel Hill) for many fruitful discussions about polyampholytes. We also thank Professor Gregory N. Parsons and Professor Stefan Franzen (NCSU) for allowing us to use their SFM and FTIR set-ups. The authors would like to thank Ms. Jan Singhass (NCSU glass shop) for her assistance in building the polymerization apparatus.

References

1. Granick S, Kumar SK, Amis EJ, Antonietti M, Balazs AC, Chakraborty AK, Grest GS, Hawker C, Janmey P, Kramer EJ, Nuzzo R, Russell TP, Safinya CR (2003) J Poly Sci Part B Poly Phys 41:2755
2. Russell TP (2002) Science 297:5583
3. Lin Q, Pearson R, Hedrick J (eds) (2004) Polymers for Microelectronics and Nanoelectronics. C.H.I.P.S, Weimar
4. Yerushalmi-Rozen R, Klein J, Fetters LJ (1994) Science 263:793
5. Gupta SA, Cochran HD, Cummings PT (1997) J Chem Phys 107:10316
6. Napper DH (1983) Steric Stabilization of Colloidal Particles. Academic Press, New York
7. Leckband D, Sheth S, Halperin A (1999) J Biomater Sci Polymer Edn 10:1125
8. Halperin A (1999) Langmuir 15:2525
9. Satulovsky J, Carignano MA, Szleifer I (2000) Proc Nat Acad Sci 97:9037

10. Nath N, Chilkoti A (2003) Adv Mater 14:1243
11. von Werne T, Patten TE (1999) J Am Chem Soc 121:7409
12. von Werne T, Patten TE (2001) J Am Chem Soc 123:7497
13. Zhao B, Brittain WJ (2000) Prog Polym Sci 25:677
14. Douglas JF, Kent MS, Satija SK, Karim A (2001) Polymer Brushes: Structure and Dynamics. In: Buschow KHJ (ed) Encyclopedia of Materials: Science and Technology. Elsevier, Amsterdam
15. Alexander S (1977) J Phys France 38:983
16. Halperin A, Tirrell M, Lodge TP (1991) Adv Polym Sci 100:31
17. Singh C, Picket GT, Balazs AC (1996) Macromolecules 29 7559
18. Currie EPK, Norde W, Cohen Stuart MA (2003) Adv Colloid Inter Sci 100–102:205
19. Meredith JC, Karim A, Amis EJ (2002) MRS Bull 27:330
20. Hoogenboom R, Meir MAR, Schubert US (2003) Macromol Rapid Commun 24:15
21. Genzer J (2002) Molecular gradients: Formation and applications in soft condensed matter science. In: Buschow KHJ, Cahn RW, Flemings MC, Ilschner B, Kramer EJ, Mahajan S (eds) Encyclopedia of Materials Science. Elsevier, Amsterdam
22. Lee JH, Khang G, Lee JW, Lee HB (1998) J Biomed Mater Res 40:180
23. Lee JH, Lee JW, Khang G, Lee HB (1997) Biomaterials 18:351
24. Jeong BJ, Lee JH, Lee HB (1996) J Colloid Inter Sci 178:757
25. Lee JH, Jeong BJ, Lee HB (1997) J Biomed Mater Res 34:105
26. Ionov L, Zdyrko B, Sidorenko A, Minko S, Klep V, Luzinov I, Stamm M (2004) Macromol Rapid Commun 25:360
27. Ionov L, Houbenov N, Sidorenko A, Minko S, Stamm M (2004) Polym Mater Sci Eng 90:104
28. Edmondson S, Osborne VL, Huck WTS (2004) Chem Soc Rev 33:14
29. Jordan R, Ulman A, Rafailovich M, Sokolov J (1999) J Am Chem Soc 121:1016
30. Advincula RC, Zhou QG, Park M, Wang SG, Mays J, Sakellariou G, Pispas S, Hadjichristidis N (2002) Langmuir 18:8672
31. Zhao B, Brittain WJ (2000) Macromolecules 33:342
32. Jordan R, Ulman A (1998) J Am Chem Soc 120:243
33. Husseman M, Mecerreyes D, Hawker CJ, Hedrick JL, Shah R, Abbott NL (1999) Angew Chem Int Ed 38:647
34. Choi IS, Langer R (2001) Macromolecules 34:5361
35. Kim NY, Jeon NL, Choi IS, Takami S, Harada Y, Finnie KR, Girolami GS, Nuzzo RG, Whitesides GM, Laibinis PE (2000) Macromolecules 33:2793
36. Juang A, Scherman OA, Grubbs RH, Lewis NS (2001) Langmuir 17:1321
37. Husseman M, Malmstrom EE, McNamara M, Mate M, Mecerreyes D, Benoit DG, Hedrick JL, Mansky P, Huang E, Russell TP, Hawker CJ (1999) Macromolecules 32:1424
38. Baum M, Brittain WJ (2002) Macromolecules 35:610
39. Pyun J, Kowalewski T, Matyjaszewski K (2003) Macromol Rapid Commun 24:1043
40. Ejaz M, Yamamoto S, Ohno K, Tsujii Y, Fukuda T (1998) Macromolecules 31:5934
41. Coessens V, Pintauer T, Matyjaszewski K (2001) Prog Polym Sci 26:337
42. Matyjaszewski K, Xia J (2001) Chem Rev 101:2921
43. Jones DM, Huck WTS (2001) Adv Mater 13:1256
44. Juang W, Kin JB, Bruening ML, Baker GL (2002) Macromolecules 25:1175
45. Wang XS, Laxcelles SF, Jackson RA, Armes SP (1999) Chem Commun 1817
46. Robinson KL, Khan MA, de Par Banez MV, Wang XS, Armes SP (2001) Macromolecules 34:3155
47. Chen G, Ito Y (2001) Biomaterials 22:2453

48. Liu H, Ito Y (2003) J Biomed Mater Res 67A:1424
49. Hypolite CL, McLernon TL, Adams DN, Chapman KE, Herbert CB, Huang CC, Distefano MD, Hu WS (1997) Bioconjugate Chem 8:658
50. Matyjaszewski K, Ziegler MJ, Arehart SV, Greszta D, Pakula T (2000) J Phys Org Chem 12:775
51. Matyjaszewski K, Greszta D, Pakula T (1997) Poly Prepr 38:709
52. Lee SB, Russell AJ, Matyjaszewski K (2003) Biomacromolecules 4:1386
53. Davis KA, Matyjaszewski K (2002) Adv Poly Sci 159:1
54. Patten TE, Matyjaszewski K (1998) Adv Mater 10:901
55. Börner HG, Duran D, Matyjaszewski K, da Silva M, Sheiko SS (2003) Macromolecules 35:3387
56. Noy A, Vezenov DV, Lieber CM (1997) Ann Rev Mater Sci 27:381
57. Ulman A (1991) An Introduction to Ultrathin Organic Films from Langmuir–Blodgett to Self Assembly. Academic Press, New York
58. Ruardy TG, Schakenraad JM, van der Mei HC, Busscher HJ (1997) Surf Sci Rep 29:1
59. Chou NJ (1990) In: Tong H-M, Nguyen LT (eds) New Characterization Techniques for Thin Polymer Films. Wiley, New York, p 289
60. Stöhr J (1992) NEXAFS Spectroscopy. Springer, Berlin Heidelberg New York
61. Genzer J, Fischer DA, Efimenko K (2003) Appl Phys Lett 82:266
62. Fischer DA, Efimenko K, Bhat RR, Sambasivan S, Genzer J (2004) Macromol Rapid Commun 25:141
63. Pohl DW (1990) In: Sheppard CJR, Mulvey T (eds) Advances in Optical and Electron Microscopy. Academic Press, London
64. Jin G, Jansson R, Arwin H (1996) Rev Sci Instr 67:2930
65. Anders S, Padmore HA, Duarte RM, Renner T, Stammler T, Scholl A, Scheinfein MR, Stohr J, Seve L, Sinkovic B (1999) Rev Sci Instr 70:3973
66. Wu T, Efimenko K, Genzer J (2002) J Am Chem Soc 124:9394
67. Chaudhury MK, Whitesides GM (1992) Science 256:1539
68. The NEXAFS experiments were carried out on the U7A NIST/Dow Materials Soft X-ray Materials Characterization Facility at the National Synchrotron Light Source at Brookhaven National Laboratory (NSLS BNL)
69. Chaudhury MK, Owen MJ (1993) J Phys Chem 97:5722
70. Allara DL, Parikh AN, Judge E (1994) J Chem Phys 100:1761
71. Huang X, Doneski LJ, Wirth MJ (1998) Anal Chem 70:4023
72. Huang X, Wirth MJ (1999) Macromolecules 32:1694
73. Wu T, Efimenko K, Genzer J (2001) Macromolecules 34:684
74. Yamamoto S, Ejaz M, Ohno K, Tsujii Y, Fukuda T (2000) Macromolecules 33:5608
75. Jones DM, Brown AA, Huck WTS (2002) Langmuir 18:1265
76. Wu T, Efimenko K, Vlček P, Šubr V, Genzer J (2003) Macromolecules 36:2448
77. In order to measure the contact angle of DI water on pure PAAm, we have first grown PAAm from a substrate covered homogeneously with the CMPE initiator. The polymerization conditions were the same as for the sample described in the text. The contact angle was determined using the technique described in the text
78. Kent MS (2000) Macromol Rapid Commun 21:243 and references therein
79. Carignano MA, Szleifer I (1995) Macromolecules 28:3197
80. Szleifer I (1996) Curr Opin Colloid Interface Sci 1:416
81. Szleifer I, Carignano MA (1996) Adv Chem Phys XCIV:165
82. Pincus P (1991) Macromolecules 24:2912
83. Zhulina EB, Birshtein TM, Borisov OV (1995) Macromolecules 28:1491
84. Zhulina EB, Borisov OV (1997) J Chem Phys 107:5952

85. Israëls R, Scheutjens JMHM, Fleer GJ (1994) Macromolecules 27:3087
86. Israëls R, Leermakers FAM, Fleer GJ, Zhulina EB (1994) Macromolecules 27:3249
87. Fleer GJ (1996) Ber Bunsenges Phys Chem 100:936
88. Israëls R, Leermakers FAM, Fleer GJ (1995) Macromolecules 28:1626
89. Wu T, Genzer J, Gong P, Szleifer I, Vlček P, Šubr V (2004) In: Advincula RC, Brittain WJ, Rühe J, Caster K (ed) Polymer Brush. Wiley, New York
90. Currie EPK, Sieval AB, Fleer GJ, Cohen Stuart MA (1999) Langmuir 15:7116
91. Currie EPK, Sieval AB, Fleer GJ, Cohen Stuart MA (2000) Langmuir 16:8324
92. Matyjaszewski K, Miller PJ, Shukla N, Immaraporn B, Gelman A, Luokala BB, Siclovan TM, Kickelbick G, Vallant T, Hoffmann H, Pakula T (1999) Macromolecules 32:8716
93. Davis KA, Matyjaszewski K (2000) Macromolecules 33:4039
94. Tomlinson MR, Genzer J (2003) Macromolecules 36:3449
95. Matyjaszewski K, Patten TE, Xia JH (1997) J Am Chem Soc 119:674
96. Kim JB, Huang WX, Bruening ML, Baker GL (2002) Macromolecules 35:5410
97. Osborne VL, Jones DM, Huck WTS (2002) Chem Commun 1838
98. Tomlinson MR, Genzer J (2003) Chem Commun 1350
99. Dautzenberg H, Jaeger W, Kotz J, Philipp B, Seidel C (1994) Polyelectrolytes. Hanser Publishers, Munich
100. Everaers R, Johner A, Joanny JF (1997) Europhys Lett 37:275
101. Harada A, Kataoka K (1999) Science 283:65
102. Walter H, Harrats C, Muller-Buschbaum P, Jerome R, Stamm M (1999) Langmuir 15:1260
103. Mahltig B, Jerome R, Stamm M (1999) Phys Chem Chem Phys 1:3853
104. Gohy JF, Creutz S, Garcia M, Mahltig B, Stamm M, Jerome R (2000) Macromolecules 33:6378
105. Bruining MJ, Blaauwgeers HGT, Kuijer R, Pels E, Nuijts RMMA, Koole LH (2000) Biomaterials 21:595
106. Hilborn J, Gupta B, Garamszegi L, Laurent A, Plummer CJG, Bisson I, Frey P, Hedrick JL (2000) Mat Res Soc Symp Proc 629:FF4.1.1
107. Chun MK, Cho CS, Choi HK (2000) J Appl Poly Sci 79:1525
108. We note that charge dissociation behavior of surface-grafted molecules can be very different from that of free molecules [Bain CD, Whitesides GM (1989) Langmuir 5:1370; van der Vegte EW, Hadziioannou H (1997) J Phys Chem B 101:9563]. However, the bulk values can still be used to obtain a good estimate of behavior in grafted polyelectrolyte systems
109. Zhang X, Xia J, Matyjaszewski K (1998) Macromolecules 31:5167
110. Zhang X, Matyjaszewski K (1999) Macromolecules 32:1763
111. Wu T (2003) PhD Thesis. NC State University
112. Tomkins HG, McGahan WA (1999) Spectroscopic ellipsometry and reflectometry: A user's guide. Wiley, New York
113. http://www.woollam.com (see tutorials)
114. Prucker O, Rühe J (1998) Langmuir 14:6893
115. Prucker O, Rühe J (1998) Macromolecules 31:592
116. Biesalski M, Rühe J (2002) Macromolecules 35:499
117. Bhat RR, Tomlinson MR, Genzer J (2004) Macromol Rapid Commun 25:270
118. Bhat RR, Tomlinson MR, Genzer J (2005) J Poly Sci Poly Phys 43:3384
119. (1991) Special issue in Science 254:1277
120. Rao CNR, Cheetham AK (2001) J Mater Chem 11:2887
121. Wohltjen H, Snow AW (1998) Anal Chem 70:2856

122. Oregan B, Gratzel M (1991) Nature 353:737
123. Nie S, Emory SR (1997) Science 275:1102
124. Colvin VL, Schlamp MC, Alivisatos AP (1994) Nature 370:354
125. Shipway AN, Katz E, Willner I (2000) Chem Phys Chem 1:18
126. Link S, El-Sayed MA (2000) Inter Rev Phys Chem 19:409
127. Alivisatos AP (1996) Science 271:933
128. Godovski DY (1995) Adv Pol Sci 119:79
129. Brus L (1997) Curr Opin Colloid Interface Sci 1:197
130. Weller H (1993) Angew Chem Int Ed Engl 105:43
131. Grabar KC, Smith PC, Musick MD, Davis JA, Walter DG, Jackson MA, Guthrie AP, Natan MJ (1996) J Am Chem Soc 118:1148
132. Grabar KC, Freeman RG, Hommer MB, Natan MJ (1995) Anal Chem 67:735
133. Kotov NA, Dekany I, Fendler JH (1995) J Phys Chem 99:13065
134. Bhat RR, Fischer DA, Genzer J (2002) Langmuir 18:3404
135. Sohn BH, Cohen RE (1997) Chem Mater 9:264
136. Hashimoto T, Harada M, Sakamoto N (1999) Macromolecules 32:6867
137. Gage RA, Currie EPK, Stuart MAC (2001) Macromolecules 34:5078
138. Bhat RR, Genzer J (2005) Appl Surf Sci, in press
139. Liu Z, Pappacena K, Cerise J, Kim JU, Durning CJ, O'Shaughnessy B, Levicky R (2002) Nano Lett 2:219
140. Bhat RR, Genzer J, Chaney BN, Sugg HW, Liebmann-Vinson A (2003) Nanotechnology 14:1145
141. Bhat RR, Genzer J (2005) Surf Sci 596:187
142. Frens G (1973) Nature Phys Sci 241:20
143. Xiao D, Wirth MJ (2002) Macromolecules 35:2919
144. Kim JU, O'Shaughnessy B (2002) Phys Rev Lett 89:238301
145. Musick MD, Keating CD, Keefe MH, Natan MJ (1997) Chem Mater 16:7825
146. Chandrasekharan N, Kamat PV (2001) Nano Lett 1:67
147. Jana NR, Gearheart L, Murphy CJ (2001) Langmuir 17:6782
148. Currie EPK, Van der Gucht J, Borisov OV, Stuart MAC (1999) Pure Appl Chem 71:1227
149. Currie EPK, Cohen Stuart MA, Fleer GJ, Borisov OV (2000) Eur Phys J E 1:27
150. McPherson T, Kidane A, Szleifer I, Park K (1998) Langmuir 14:176
151. Carignano MA, Szleifer I (2000) Biointerfaces 18:169
152. Szleifer I (1997) Physica A 244:370
153. Szleifer I (1997) Biophys J 72:595
154. Fang F, Szleifer I (2001) Biophys J 80:2568
155. Fang F, Szleifer I (2002) Langmuir 18:5497
156. Halperin A, Leckband DE (2000) C R Acad Sci Paris t 1 Série IV:1171
157. Efremova NV, Sheth SR, Leckband DE (2001) Langmuir 17:7628
158. Zhu B, Eurell T, Gunawan R, Leckband D (2001) J Biomed Mater Res 56:406
159. Sheth SR, Efremova N, Leckband DE (2000) J Phys Chem B 104:7652
160. Gombotz WR, Guanghui W, Horbett TA, Hoffman AS (1991) J Biomed Mater Res 25:1546
161. Rouce FR Jr, Ratner BD, Horbett TA (1982) In: Cooper SL, Peppas NA (eds) Biomaterials: Interfacial Phenomena and Applications. American Chemical Society, Washington, DC
162. Bohnert JL, Horbett TA, Ratner BD, Royce FH (1988) Investig Opth Vis Sci 29:362
163. Morra M, Cassinelli C (1999) Langmuir 15:4658
164. Bhat RR, Genzer J (2004) Mat Res Soc Symp Proc 804:J5.8.1

165. Bhat RR, Chaney BN, Rowley J, Liebmann-Vinson A, Genzer J (2005) Adv Mater 17:2802
166. (2003) Special issue of Macromol Rapid Commun, Volume 24, Issue 1
167. (2004) Special issue of Macromol Rapid Commun, Volume 25, Issue 1
168. Zhao H, Beysens D (1995) Langmuir 11:627
169. Genzer J, Kramer EJ (1998) Erophys Lett 44:180
170. Smith AP, Douglas JF, Meredith JC, Amis EJ, Karim A (2001) Phys Rev Lett 87:Art No 015503
171. Smith AP, Sehgal A, Douglas JF, Karim A, Amis EJ (2003) Macromol Rapid Commun 21:131
172. Meredith JC, Sormana JL, Keselowsky BG, Garcia AJ, Tona A, Karim A, Amis EJ (2003) J Biomed Mat Res 66A:483
173. Lord SJ, Sheiko SS, LaRue I, Lee H-I, Matyjaszewski K (2004) Macromolecules 37:4235

Adv Polym Sci (2006) 198: 125–147
DOI 10.1007/12_061
© Springer-Verlag Berlin Heidelberg 2005
Published online: 20 December 2005

Surface Rearrangement of Diblock Copolymer Brushes—Stimuli Responsive Films

William J. Brittain[1] (✉) · Stephen G. Boyes[1,2] · Anthony M. Granville[1] ·
Marina Baum[1] · Brian K. Mirous[1] · Bulent Akgun[1] · Bin Zhao[1,3] ·
Carl Blickle[1] · Mark D. Foster[1]

[1]Department of Polymer Science, The University of Akron, Akron, OH 44325-3909, USA
wjbritt@uakron.edu

[2]Department of Polymer Science, The University of Southern Mississippi,
Hattiesburg, MS 39406, USA

[3]Department of Chemistry, The University of Tennessee, Knoxville, TN 37996, USA

Abstract This article reviews results from our group of the synthesis and characterization of diblock copolymer brushes. Results from the literature are also covered. We report a wide variety of diblock compositions and compare the miscibility of the two blocks with the tendency to rearrange in response to block-selective solvents. Also, we describe the types of polymerization methods that can be utilized to prepare diblock copolymer brushes. We have compared the molecular weight of free polymer and the polymer brush based on results from our laboratory and other research groups; we have concluded that the molecular weight of the free polymer and that of degrafted polymer brushes is similar.

Keywords ATRP · Block copolymers · Polymer brushes · Stimuli-responsive · Thin films

Abbreviations

ATRP	atom transfer radical polymerization
RAFT	reversible addition fragmentation transfer polymerization
PS	polystyrene

PMMA   poly(methyl methacrylate)
PMA   poly(methyl acrylate)
PDMA   poly(N,N-dimethylacrylamide)
SAM   self-assembled monolayer
PDMAEMA   poly((N,N'-dimethylamino)ethyl methacrylate)
RATRP   reverse atom transfer radical polymerization
PAA   poly(acrylic acid)
PFS   poly(pentafluorostyrene)
PHFA   poly(heptadecafluorodecyl acrylate)
PTFA   poly(trifluoroethyl acrylate)
XPS   X-ray photoelectron spectroscopy
ATR-FTIR   attenuated total reflectance Fourier transform infrared spectroscopy
TGA   thermal gravimetric analysis
PDI   polydispersity index
AFM   atomic force microscopy

1
Introduction

Polymer brushes refer to an assembly of polymer chains which are tethered by one end to a surface or interface [1,2]. Tethering of the chains in close proximity to each other forces the chains to stretch away from the surface to avoid overlapping. Polymer brushes are typically synthesized by two different methods, physisorption and covalent attachment. Of these methods, covalent attachment is preferred as it overcomes the disadvantages of physisorption which include thermal and solvolytic instabilities [3,4]. Covalent attachment of polymer brushes can be achieved by either "grafting to" or "grafting from" techniques. The grafting-to technique involves tethering preformed end-functionalized polymer chains to a suitable substrate [5]. This technique often leads to low grafting density and low film thickness, as the polymer molecules must diffuse through the existing polymer film to reach the reactive sites on the surface. The steric hindrance for surface attachment increases as the tethered polymer film thickness increases. To overcome this problem, the grafting-from approach can be used and has generally become the most attractive way to prepare thick, covalently tethered polymer brushes with a high grafting density [3]. The grafting-from technique involves the immobilizing of initiators onto the substrate followed by in situ surface-initiated polymerization to generate the tethered polymer brush. Surface immobilized initiators are usually generated using self-assembled monolayer (SAM) techniques [6,7]. As the chains are growing from the surface, the only limit to propagation is diffusion of monomer to the chain ends, thus resulting in thick tethered polymer brushes with high grafting density.

The field of stimuli-responsive films has grown enormously in the past few years. This review will principally concentrate on work from the primary author's research, but it is obligatory and responsible to highlight some of the recent advances in other research groups. While not comprehensive, this succinct overview will help the reader in the identification of important and relevant research in other groups.

Recent advances in polymer synthesis techniques have given rise to the importance of controlled/"living" free radical polymerization, as it provides a number of advantages over traditional free radical techniques [8]. Although other polymerization methods have been used, one of the main advantages controlled/living free radical polymerization provides for polymer brush synthesis is control over the brush thickness, via control of molecular weight and narrow polydispersities [9, 10]. Another advantage that a controlled/living free radical system provides is the ability to produce polymer brushes of specific architectures and large range of polymerizable monomers.

Without attempting a bias towards our own work on stimuli-responsive diblock copolymer brushes, the group research of Minko, Stamm, Tsukruk and Luzinov have been equally or more influential in the field of stimuli-responsive films. One important fact to consider throughout any discussion of covalently attached polymer chains is the distinction between chains which exist in the "brush" regime vs. the "mushroom" regime. How this affects stimuli-responsive properties remains an open question; but, for each polymer system, there is clearly a grafting density that defines a transition from the mushroom to brush regime. The reader is referred to a particularly important paper by Genzer and co-workers [11] who clearly defined a grafting density that represented this transition. Although their study only concentrated on polyacrylamide brushes, it serves as a general guide to an approximate grafting density at which an investigator can claim that their covalently attached polymer exists in the brush regime. One clear assertion that can be drawn from the literature is that any grafted polymer prepared by the grafting-to technique will inevitably lead to a covalently attached system in the mushroom regime which will be characterized by low film thicknesses.

Stimuli-responsive surface polymers are typically characterized by either binary systems (two chemically dissimilar chains attached to the same surface) or homopolymer brushes based on poly(acrylamide) [especially poly(N-isopropyl)acrylamide, pNIPAA] which demonstrate temperature-dependent conformational changes. Desai and co-workers [12] performed ATRP of N-isopropyl acrylamide (NIPAA) onto oxidized films of oxidized polypropylene and observed stimuli-responsive behavior. Scanning-probe lithography was used by Zauscher et al. [13, 14] to create a patterned array of ATRP initiators for the preparation of nanopatterned pNIPAA polymer chains that demonstrated reversible height changes via a transition cycling of the substrate between water and water/methanol mixtures. Grafting-from of NIPAA from polystyrene latex particles was reported by Kizhakkedathu and co-workers [15]; the hydrodynamic thickness of the latex particles was controlled by the polymerization conditions and block copolymers with N,N-dimethylacrylamide were also reported. Farhan and Huck [16] reported the formation of pNIPAA layers on oxidized polyester films; in addition, they created patterned thermoresponsive films using microcontact printing. One interesting report was the use of pH to control film behavior in a modified copolymer of NIPAA and glycinylacrylamide [17]. Zhu et al. [17] observed typical polyelectrolyte brush behavior above pH 8.0 but the film thickness decreased substantially at lower pH values (5.0) presumably due to hydrogen-bonding between glycine side chains. Using a Si(100) (Si – H) surface for a grafting-to method, Xu and co-workers [18] studied the cell attachment of the cell line 3T3 Swiss albino on both homopolymers of pNIPAA and poly(ethylene glycol) monomethacrylate (PEGMA). Above the lower critical solution temperature (LCST) of pNIPAA, the cells proliferated but below the LCST, cells detached spontaneously. The PEGMA surface was very effective at preventing cell attachment and growth. However, incorporation of PEGMA units into the pNIPAA chains via copolymerization resulted in a more rapid cell detachment during the LCST transition. This work represents an interesting example of a stimuli-responsive surface for the controlled adhesion of cells. Somewhat related to the work on pNIPAA brushes is work from the group of Ito and co-workers [19] where they have studied the permeability of membranes modified with poly(acrylic acid) as a function of pH. Although Ryan and co-workers [20] did not focus on homopolymer brushes, they studied innovative multiblock systems composed of a hydrophobic end-block and either polyacid or polybase mid-block; this comprehensive study used a battery of techniques to follow molecular shape change and macroscopic deformation.

There have been a number of reports on binary brushes which refer to a mixture of two different polymer chains attached to the same surface. Zhao and co-workers [21, 22] have reported some novel systems using a surface immobilized dual initiator which clearly produces polymer brushes. Their work will be discussed later in this review. An extensive body of work has been published on binary systems using predominately grafting-to methods; many of these reports have provided provocative results on stimuli-responsive polymer systems. Minko, Stamm and co-workers [23] demonstrated a reversible patterning of a stimuli-responsive film based on a binary brush composed of poly(2-vinylpyridine) (PVP) and polyisoprene. Exposure to different solvents demonstrated a switching behavior; furthermore, crosslinking of the polyisoprene by illumination through masks created a patterned surface that showed location-dependent stimuli-responsive behavior. In related work, switchable binary brushes were prepared on silicon wafers composed of PVP and a polystyrene copolymer containing a photodimerizing phenylindene component [24]. Water contact angle changes of 40 degrees were observed depending on the polarity of the solvent to which the sample was exposed. The structural state of the switchable surface could be fixed by photoinduced crosslinking of the phenylindene unit. In subsequent work, the same principle authors developed an "anchoring layer" to prepare binary brushes [25]. They compared the grafting of polystyrene (PS) and PVP end-grafted chains to either an epoxysilane-modified monolayer or a macromolecular layer composed of poly(glycidyl methacrylate). Sequential deposition of PS and PVP produced binary brushes that differed in wettability and nanomorphology. Minko, Stamm and co-workers [26] fabricated binary films of PS and PVP on polyamide polymer surfaces that had been first treated in an ammonia plasma followed by immobilization of an azo initiator. They observed significant differences in grafting-from a silicon surface and the derivatized polyamide surface. They extended this technique to fabric modification and reported water contact angles up to 150 degrees. Recently, two excellent reviews of the work from Minko and co-workers [27, 28] have been published which details the work described above plus other efforts from their group and their collaborators.

1.1
Block Copolymer Brushes

One of the most interesting of these architectures produced to date are block copolymer brushes. Block copolymer brushes are interesting due to the fact that vertical phase separation results when the block copolymer chains are tethered by one end to a surface or substrate. By changing the grafting density, chain length, relative block length, composition of the blocks or the interaction energy between the blocks and the surrounding environment, the formation of a variety of novel well-ordered structures have been predicted by theory [29, 30] and in some cases demonstrated experimentally [21, 31–34].

2
Synthesis of Block Copolymer Brushes

2.1
Results from Other Research Groups

In this review, synthesis of block copolymer brushes will be limited to the grafting-from method. Hussemann and coworkers [35] were one of the first groups to report copolymer brushes. They prepared the brushes on silicate substrates using surface-initiated TEMPO-mediated radical polymerization. However, the copolymer brushes were not diblock copolymer brushes in a strict definition. The first block was PS, while the second block was a 1 : 1 random copolymer of styrene/MMA. Another early report was that of Matyjaszewski and coworkers [36] who reported the synthesis of poly(styrene-*b*-*tert*-butyl acrylate) brushes by atom transfer radical polymerization (ATRP).

This was the first report using ATRP and sequential monomer addition. Hydrolysis of these diblock copolymer brushes yielded poly(styrene-b-acrylic acid) brushes.

During the last 5 years, there have been several reports of multiblock copolymer brushes by the grafting-from method. The most common substrates are gold and silicon oxide layers; but there have been reports of diblock brush formation on clay surfaces [37] and silicon-hydride surfaces [38]. Most of the newer reports have utilized ATRP [34, 38–43] but there have been a couple of reports that utilized anionic polymerization [44, 45]. Zhao and co-workers [21, 22] have used a combination of ATRP and nitroxide-mediated polymerization to prepare mixed poly(methyl methacrylate) (PMMA)/polystyrene (PS) brushes from a difunctional initiator. These Y-shaped brushes could be considered block copolymers that are surface immobilized at the block junction.

2.2
Synthesis of Block Copolymer Brushes in Our Group

The first diblock copolymer brushes synthesized in our group were made by a combination of carbocationic polymerization and ATRP (Scheme 1) [46]. Zhao and co-workers [47] synthesized diblock copolymer brushes consisting of a tethered chlorine-terminated PS block, produced using carbocationic polymerization, on top of which was added a block of either PMMA, poly(methyl acrylate) (PMA) or poly((N,N′-dimethylamino)ethyl methacrylate) (PDMAEMA), synthesized using ATRP. The thickness of the outer poly(meth)acrylate block was controlled by adding varying amounts of free initiator to the ATRP media. It has been reported that the addition of free initiator is required to provide a sufficiently high concentration of deactivator, which is necessary for controlled polymerizations from the sur-

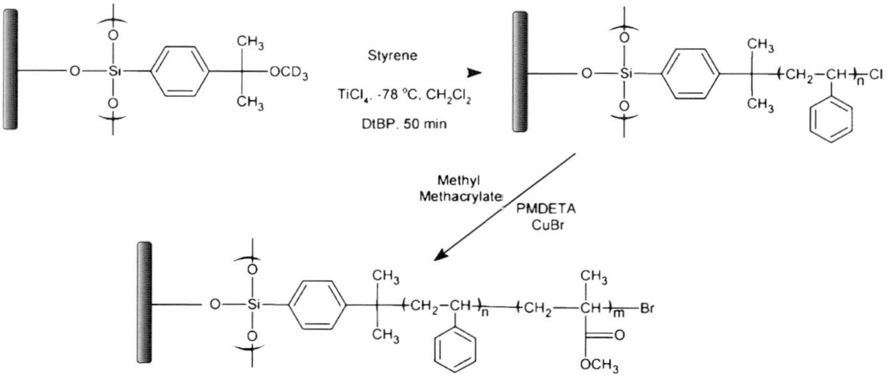

Scheme 1 Synthesis of surface-immobilized diblock copolymer brush (Si/SiO$_2$//PS-b-PMMA) using a combination of carbocationic polymerization and ATRP

face [35]. Table 1 summarizes the properties of some of the diblock copolymer brushes.

The first diblock copolymer brush synthesized completely using controlled/living free radical polymerization techniques in our group was by Sedjo and co-workers [48]. In this work a tethered diblock copolymer of PS and PMMA was synthesized using a combination of reverse atom transfer radical polymerization (RATRP) and standard ATRP techniques (Scheme 2). The properties of this diblock copolymer brush can be seen in Table 1. RATRP involves initiation by conventional radical initiators in the presence of an ATRP deactivator. RATRP has been shown to produce polymers that are end-functionalized with a transferable halogen, thus allowing continued polymerization [49, 50]. To perform RATRP from the surface, an azo-initiator was first immobilized on the silicon substrate followed by the polymerization of styrene in the presence of copper(II)bromide and ligand. This resulted in the formation of a tethered block of PS with a terminal bromine group. The terminal bromine group was subsequently used to initiate MMA under standard ATRP conditions.

Table 1 Summary of the properties of diblock copolymer brushes

Diblock copolymer brush structure[a]	Thickness of tethered block [b] (nm)	Thickness of outer block[b] (nm)	Polymerization technique[c]	Refs.
Si/SiO₂//PS-b-PMMA	28	11	Cationic/ATRP	[46, 47]
Si/SiO₂//PS-b-PMA	24	9	Cationic/ATRP	[47]
Si/SiO₂//PS-b-PDMAEMA	27	3	Cationic/ATRP	[47]
Si/SiO₂//PS-b-PMMA	25	7	RATRP/ATRP	[48]
Si/SiO₂//PS-b-PDMA	11	12	RAFT	[51]
Si/SiO₂//PDMA-b-PMMA	11	10	RAFT	[51]
Si/SiO₂//PS-b-P(t-BA)	21	17	ATRP	[52]
Si/SiO₂//PS-b-PAA	21	8	ATRP/Hydrolysis	[52]
Si/SiO₂//PMA-b-P(t-BA)	14	16	ATRP	[52]
Si/SiO₂//PMA-b-PAA	14	9	ATRP/Hydrolysis	[52]
Si/SiO₂//PS-b-PPFS	16	5	ATRP	[53]
Si/SiO₂//PS-b-PHFA	10	6	ATRP	[53]
Si/SiO₂//PMA-b-PPFS	11	5	ATRP	[53]
Si/SiO₂//PMA-b-PHFA	15	5	ATRP	[53]

[a] PS—polystyrene, PMMA—poly(methyl methacrylate), PMA—poly(methyl acrylate), PDMAEMA—poly((N,N-dimethylamino)ethyl methacrylate), PDMA—poly(dimethylacrylamide), P(t-BA)—poly(tert-butyl acrylate), PAA—poly(acrylic acid), PPFS—poly(pentafluorostyrene), PHFA—poly(heptadecafluorodecyl acrylate)
[b] Representative structure is Si/SiO₂//tethered block-b-outer block
[c] ATRP—atom transfer radical polymerization, RATRP—reverse atom transfer radical polymerization, RAFT—reversible addition fragmentation transfer polymerization

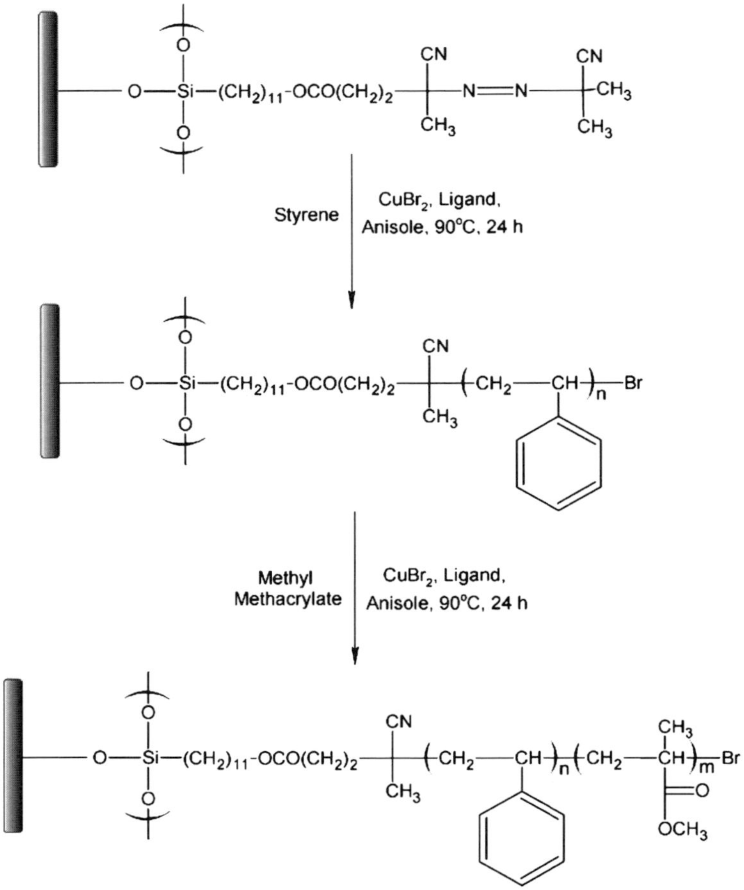

Scheme 2 Synthesis of surface-immobilized diblock copolymer brush (Si/SiO₂//PS-*b*-PMMA) using reverse atom transfer radical polymerization and ATRP

To make further use of the azo-initiator, tethered diblock copolymers were prepared using reversible addition fragmentation transfer (RAFT) polymerization. Baum and co-workers [51] were able to make PS diblock copolymer brushes with either PMMA or poly(dimethylacrylamide) (PDMA) from a surface immobilized azo-initiator in the presence of 2-phenylprop-2-yl dithiobenzoate as a chain transfer agent (Scheme 3). The properties of the diblock copolymer brushes produced can be seen in Table 1. The addition of a "free" initiator, 2,2′-azobisisobutyronitrile (AIBN), was required in order to obtain a controlled polymerization and resulted in the formation of free polymer chains in solution.

In order to produce block copolymer brushes by ATRP directly from the surface, the ATRP initiator, (11-(2-bromo-2-methyl)propionyloxy)undecyltri-

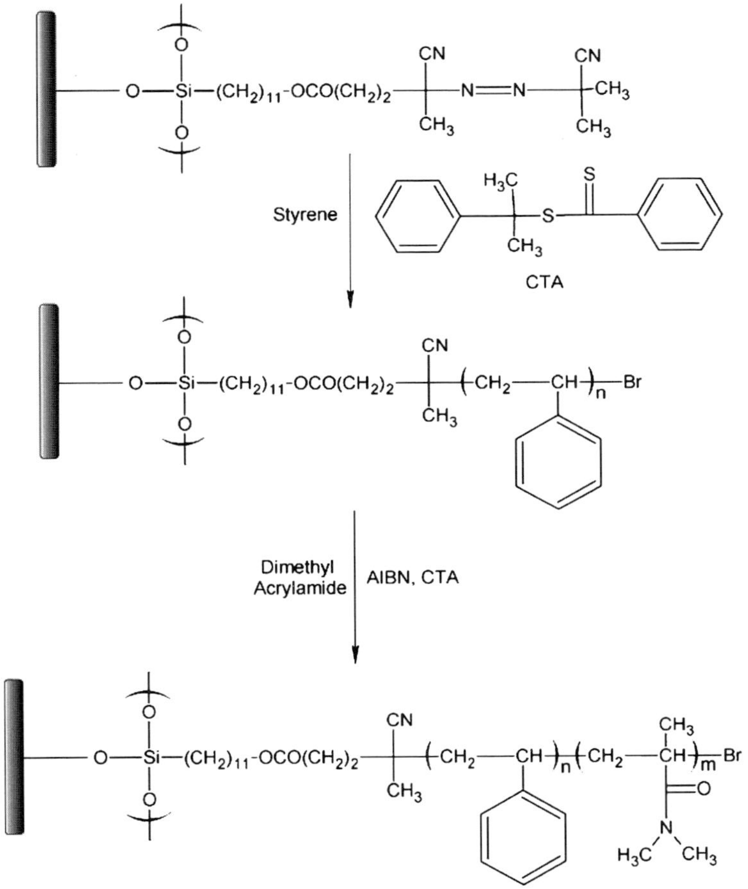

Scheme 3 Synthesis of surface-immobilized diblock copolymer brush (Si/SiO$_2$//PS-*b*-PDMA) using reverse addition fragmentation transfer polymerization

chlorosilane, was prepared and immobilized on silicon substrates. From this immobilized bromo-isobutyrate type ATRP initiator both Boyes and co-workers [52] and Granville and co-workers [53] were able to synthesize diblock copolymer brushes using ATRP. Boyes and co-workers [52] synthesized diblock copolymer brushes of either PS or PMA and poly(*tert*-butyl acrylate) (P(*t*-BA)) using ATRP, with subsequent hydrolysis of the P(*t*-BA) to poly(acrylic acid) (PAA) (Scheme 4). The properties of these diblock copolymer brushes can be seen in Table 1. The diblock copolymer brushes, Si/SiO$_2$//PS-*b*-PAA and Si/SiO$_2$//PMA-*b*-PAA, were both treated with aqueous silver acetate to produce polyelectrolyte diblock copolymer brushes. The polyelectrolyte brushes were subsequently reduced using H$_2$ resulting in the formation of silver nanoparticles within the diblock copolymer brush [52]. Granville and co-workers [53] used similar ATRP techniques to synthesize diblock copolymer brushes that contained the semifluorinated monomers pentafluorostyrene (PFS) and heptadecafluorodecyl acrylate (HFA). The properties of these diblock copolymer brushes can also be seen in Table 1. The use of fluorinated monomers to produce outer blocks of either PPFS or PHFA resulted in surfaces that were highly hydrophobic [53].

We also synthesized a triblock copolymer brush using sequential monomer addition and ATRP. Boyes and co-workers [33] synthesized the ABA type triblock copolymer brushes of PS and PMA to produce Si/SiO$_2$//PS-b-PMA-b-PS and Si/SiO$_2$//PMA-b-PS-b-PMA brushes. We observed incomplete re-initiation for the third block of Si/SiO$_2$//PS-b-PMA-b-PS (as indicated by a 3 nm thickness); the incomplete re-initiation was attributed to radical-radical termination occurring in the formation of the previous blocks. In the case of the Si/SiO$_2$//PMA-b-PS-b-PMA brush, the outer PMA block had a thickness of 15 nm, which is close to the target thickness of 20 nm, indicating that the degree of termination occurring in this system was less.

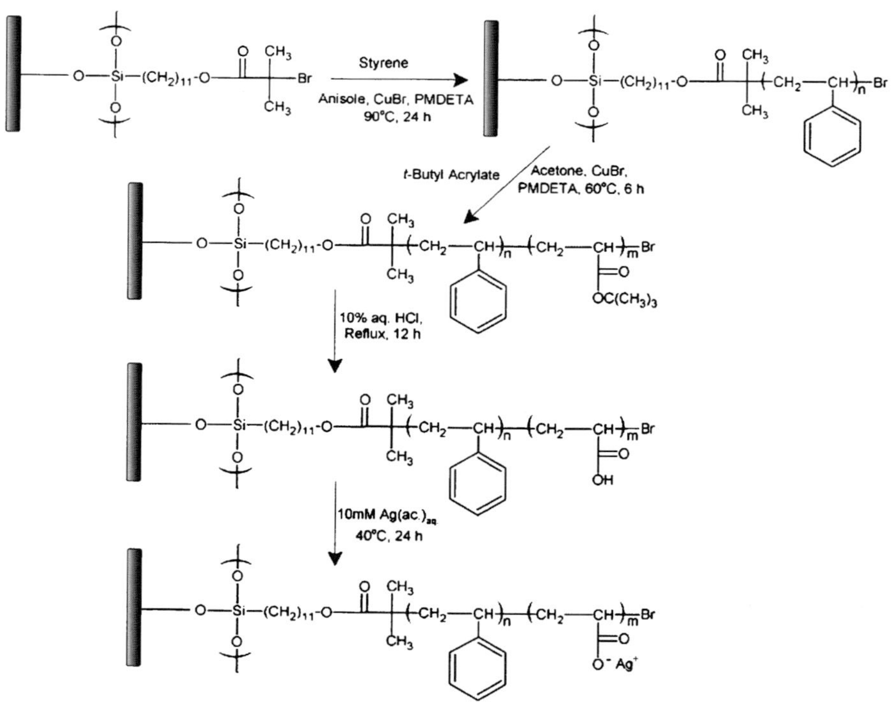

Scheme 4 Synthesis of surface-immobilized polyelectrolyte diblock copolymer brush (Si/SiO$_2$//PS-b-PAA(Ag$^+$)) using ATRP

3
Correlation of Brush Thickness with Molecular Weight

Before describing the rearrangement of diblock polymer brushes, we will re-view results from our laboratory and the literature that give a clearer picture of the structure of both homopolymer and copolymer brushes. Specifically, what is known about grafting density and the relationship between brush mo-lecular weight and film thickness. This review focuses on brushes made by either ATRP or RAFT.

Our report [51] on RAFT from surfaces appeared nearly simultaneously with the report of Tsujii and co-workers [54]. Tsujii and co-workers con-centrated on a kinetic analysis of PS homopolymer brush formation while we were more interested in using the unique feature of RAFT to polymer-ize acrylamides. We worked primarily on flat substrates and conducted RAFT by starting with an immobilized azo initiator and running the polymeriza-tion in the presence of a RAFT agent and free azo initiator (AIBN). Because free polymer (polymer not covalently bound to the surface) was formed in our RAFT studies, it was possible to compare the molecular weight of free polymer to the degrafted polymer. We used spherical silica particles and immobilized Rühe's initiator [9]. Using thermogravimetric analysis (TGA), we [55] determined that the grafting density was 0.7 initiator molecules/nm^2, a value that corresponds well with Prucker and Rühe [9] (0.8–1.6 initiator molecules/nm^2). We determined the molecular weight for both PMMA and PS brushes. For PMMA, M_n (free polymer) = 15 900 g/mol, PDI = 1.22 and M_n (degrafted polymer) = 19 200 g/mol, PDI = 1.29. For PS, M_n (free poly-mer) = 10 600 g/mol, PDI = 1.11 and M_n (degrafted polymer) = 11 400 g/mol, PDI = 1.14. There is a close correspondence between both the molecular weight and PDI. The relatively narrow PDI is consistent with a living rad-ical process. Using the molecular weight and TGA data, we determined that the initiator efficiency (f) ranged from 0.15–0.19 [55]. Tsujii and co-workers [54] published a graph comparing molecular weight for degrafted and free PS; inspection suggests that the difference in M_n values was never greater than 4000 g/mol and the PDI for degrafted polymer was slightly higher than free polymer, as we also observed. Patten et al. [56, 57] ob-served similar trends. The significance of these results is that analysis of free polymer provides a reasonable estimate of the brush molecular weight, thus eliminating the time-consuming chore of performing an analogous proced-ure on high surface area supports. Knowing the brush molecular weight is important because it allows one to calculate the occupied area of a single polymer brush chain (A_X) using the following equation: $A_X = M_n/(h\rho N_A)$ where ρ = brush bulk density, h = dry brush thickness and N_A is Avogadro's Number. Knowing the relationship between grafting density and solvent-induced rearrangement of diblock copolymer brushes is one of our key objectives.

We have repeated similar degrafting experiments for brush formation via ATRP. While there have been reports on degrafting using conventional radical polymerization [10, 58], this discussion will be limited to brush formation by ATRP. In unpublished work [59], we immobilized an ATRP initiator, (11-(2-bromo-2-methyl)propionyloxy)undecyltrichlorosilane) on Stöber silica and conducted a styrene polymerization. Degrafting of the PS brushes was conducted by etching of the silica cores with HF. From TGA analysis of the immobilized initiator and the corresponding PS brush system, we determined that there are 4.8 initiator molecules/nm^2 and $f = 0.06$. The initiator density corresponds well to the values of 2.4–5.0 reported by Patten and co-workers [56, 57] for the immobilization of (2-(4-chloromethylphenyl)ethyl)dimethylethoxysilane on a similar support.

Brush formation by ATRP can be accompanied by free polymer if the process is conducted with a free initiator. A high concentration of a deactivating Cu(II) complex is necessary for control of ATRP [36]. For ATRP from a surface, the small amount of initiator tethered to the substrate provides too low a concentration of Cu(II) to control the polymerization. One solution to this problem is to add Cu(II) at the beginning of the brush synthesis [36]. Another solution is to use a free initiator, which generates a sufficient Cu(II) concentration in situ. We prefer this latter approach because others and we have observed that the molecular weight of the free polymer roughly corresponds to the molecular weight of the polymer brush. For our PS brush silica gel experiments, [59] we observed M_n (free polymer) = 19 600 g/mol, PDI = 1.11 and M_n (degrafted polymer) = 27 100 g/mol, PDI = 1.57. Von Werne and Patten [56, 57] reported better correspondence for an analogous experiment (ATRP on spherical silica); for PMMA – M_n (free polymer) = 50 700 g/mol, PDI = 1.16 and M_n (degrafted polymer) = 57 100 g/mol, PDI = 1.26; for PS – M_n (free polymer) = 43 800 g/mol, PDI = 1.22 and M_n (degrafted polymer) = 46 300 g/mol, PDI = 1.29. Hawker and co-workers [35] performed nitroxide-mediated radical polymerization of styrene on silica gel; they observed M_n (free polymer) = 48 000 g/mol, PDI = 1.20 and M_n (degrafted polymer) = 51 000 g/mol, PDI = 1.14. In summary, these three studies have demonstrated that there is a reasonable and reliable correlation between the molecular weight and PDI for free polymer and the polymer brush for brush growth via ATRP.

Somewhat related to studies on silica gel is a report by Ejaz and co-workers [60] where they studied the ATRP of MMA on porous glass filters. They examined the relationship between the molecular weight of free polymer and degrafted polymer. While they did not provide the raw data, they presented the results in a graph, which indicates that there was a 35% or less discrepancy between the polymer covalently attached to the frit and that produced by the free initiator. The final literature report that bears notice is that from Kim, Bruening and Baker [61]. To the best of our knowledge of this writer, this is the only report that compared the molecular weight of degrafted polymer brushes grown on a flat substrate; they used a large gold substrate and degrafted the chains by treatment with I_2. Because they did not use a free initiator, a comparison between free and degrafted polymer was not possible. They observed a nonlinear dependence where the reported M_n of a 33 nm thick film was 33 100 g/mol, while $M_N = 68\,900$ g/mol is observed for a 40 nm thick film. Interestingly, the results of Kim and co-workers match those of Hawker and co-workers [35] at lower molecular weights. One interesting finding in the report by Kim and co-workers is an estimated $f = 0.10$ for ATRP from thiol/gold initiators. This value is reasonably close to the value we observed in our laboratories using ATRP from spherical silica.

In summary, upon review of results from our own laboratories combined with literature results, it is now possible to make reasonable conclusions about the molecular weight and dispersity of polymer brushes. First, it seems that f for living polymerizations approximates 0.10. Second, there is good correspondence between the M_n and PDI of free polymer and degrafted polymer.

4
Rearrangement of Block Copolymer Brushes

The behavior of tethered diblock copolymer brushes is interesting because the bottom block is highly constrained by covalent attachment to the silicate surface and localization of the other end at the diblock interface. Theoretical studies using self-consistent field calculations, scaling arguments and computer simulations have indicated that tethered block copolymer brushes exhibit complex behaviors that depend on many factors [29, 30, 62–65]. These factors include χ, overall molecular weight (N), volume fraction of one block, Kuhn length (flexibility of backbone), grafting density, environmental conditions (solvent, temperature) and the surface free energy of each block in the air. One of the most interesting structures is the "pinned micelle" structure, which can be formed when tethered AB diblock copolymer brushes are treated with a block-selective solvent [29]. To the best of our knowledge, the nomenclature of pinned micelles was originally introduced by Balazs and co-workers [29, 30] and we have adopted this terminology to describe our systems.

4.1
Nanomorphology of Si/SiO$_2$//PS-b-PMMA Brush

We reported the synthesis of Si/SiO$_2$//PS-b-poly(acrylate) tethered diblock copolymer brushes [31, 32, 46, 47]. The properties of these diblock brushes were studied using water contact angles, ellipsometry, X-ray photoelectron spectroscopy (XPS), FTIR spectroscopy and atomic force microscopy (AFM). For a sample with a 26 nm PS layer and a 9 nm PMMA layer, the advancing water contact angle increased from 75° (characteristic of PMMA) to 99° (characteristic of polystyrene) after treatment with cyclohexane; subsequent treatment with CH_2Cl_2 returned the contact angle to the original value of 75°. This contact angle change was attributed to reversible changes in the chemical composition at the polymer–air interface. XPS analysis indicated large compositional changes after treatment with CH_2Cl_2 and cyclohexane which are consistent with the contact angle observations.

For a sample with a 23 nm PS layer and 14 nm PMMA layer, AFM was used to study surface morphological changes [32]. It was found the surface is relatively smooth with a roughness of 0.77 nm after CH_2Cl_2 treatment (Fig. 1); treatment with cyclohexane at 35 °C for 1 h increased the surface roughness to 1.79 nm and created an irregular worm-like structure on the surface. A nanopattern was formed if mixed solvents of CH_2Cl_2 and cyclohexane were used and the composition was gradually changed from CH_2Cl_2 to cyclohexane (Fig. 2). The advancing water contact angle of this surface was 120°. We speculated that this nanopattern corresponds to a pinned micelle nanomorphology (Scheme 5), consistent with the theoretical predictions of Balazs and co-workers [29]. Our interpretation of this image as periodic is simply based on a visual inspection; we did not perform Fourier transform analysis to confirm the degree of periodicity.

We explored the relationship between average domain size as deduced by AFM and block lengths for a $Si/SiO_2//PS-b-PMMA$ diblock brush [31]. We assumed that the block length is proportional to ellipsometric film thickness. Table 2 contains a summary of the experimental relationship between block lengths (as determined by ellipsometry) vs. average domain diameter for the observed nanomorphologies. For the three samples studied, the largest di-

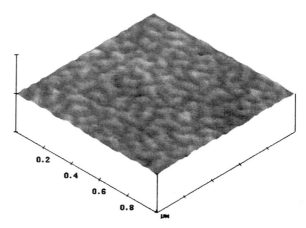

Fig. 1 AFM image of the tethered $Si/SiO_2//PS-b-PMMA$ brushes with 23 nm thick PS layer and 14 nm thick PMMA layer after treatment with dichloromethane at room temperature for 30 min and drying with a clean air stream

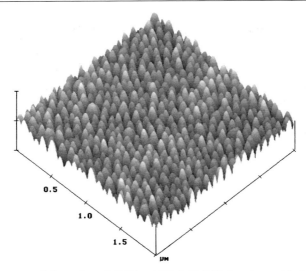

Fig. 2 AFM image of the tethered Si/SiO$_2$//PS-*b*-PMMA brushes with 23 nm thick PS layer and 14 nm thick PMMA layer after a gradual treatment with cyclohexane

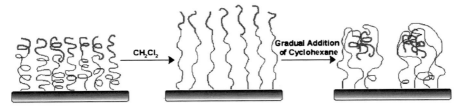

Scheme 5 Speculative model for nanopattern formation from tethered Si/SiO$_2$//PS-*b*-PMMA

Table 2 Average AFM domain diameter vs. diblock brush thickness [30]

PS Thickness, nm [a]	PMMA Thickness, nm [a]	Domain diameter, nm [b]	Roughness, nm [b]
15	3	46	5.3
23	14	85	13.1
26	17	113	11.7

[a] Thickness determined by ellipsometry
[b] Domain diameter and roughness determined by AFM

block variable is the PMMA thickness. As the thickness of the PMMA layer increases, the average domain diameter of the surface-immobilized micelles also increases. Balazs and co-workers predicted this experimental observation [29]; self-consistent field theory indicated that pinned micelles should be observed for tethered diblocks where the more soluble block (in our case, PS) is attached to the surface. Furthermore, Balazs and co-workers predicted that the size of the pinned micelles should increase with the size of the less-soluble block (in our case, PMMA), as we have observed.

These intriguing results prompted us to study the effect of different diblock brush compositions on the rearrangement. Although we successfully prepared a series of different diblock brushes and did observe changes in the surface properties (see Sect. 4.2 below), only the Si/SiO$_2$//PS-b-PMMA system displayed a periodic nanomorphology (Fig. 2). The literature contains at least two examples of block brushes that exhibit a similar, periodic nanomorphology in AFM analysis. Zhao and co-workers [21, 22] prepared a well-defined mixed PMMA/PS brush using an asymmetric difunctional initiator-terminated SAM. For mixed brushes where the PS M_n is slightly lower or similar to PMMA M_n, a periodic nanomorphology was observed in the AFM after treatment with acetic acid (a block-selective solvent for PMMA). Huang and co-workers [34] prepared a triblock brush on gold composed of PMMA-b-poly(N,N-dimethylaminoethyl methacrylate)-b-PMMA. A featureless surface was observed for the sample treated with a nonselective solvent (dichloromethane). Methanol was gradually added to the solvent until the composition was 99.5/0.5 (v/v) methanol/dichloromethane; AFM analysis of the triblock revealed a periodic nanomorphology that was attributed to surface-immobilized micelles. We also reported a triblock brush and did observe significant changes in the AFM after treatment with block-selective solvents; however, the observed features were much more disperse (Fig. 3) [33].

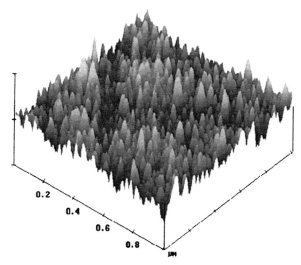

Fig. 3 AFM image of Si/SiO$_2$//PMA-b-PS-b-PMA brushes after treatment with cyclohexane

4.2
Expanded Composition of Diblock Copolymer Brushes

One important goal of this research was to expand the composition of block copolymer brushes. The motivation was to prepare dissimilar (e.g., greater disparity in hydrophobicity) blocks that might show greater changes in the surface properties before and after switching and to use blocks that would respond to nonsolvent induced stimuli (e.g., pH, ionic strength, temperature). We expanded our composition to include the monomers listed in Table 3. Also given in Table 3 are solubility parameters (δ) that were calculated using group contribution methods. The greater the difference in δ, the larger the Flory–Huggins interaction parameter (χ) and thus the less likely that two blocks would be miscible. It is logical to assume that the rearrangement of diblock copolymer brushes would be related to χ; one would predict that the diblocks with the largest χ values would be the least likely to rearrange.

Table 4 contains some representative data for solvent switching of a series of diblock brushes where the bottom block (adjacent to the silicate) is PS. Column 2 of this table contains the advancing, water contact angles for the system in the extended state; that is, the contact angle represents the composition of the upper block. Treatment of the diblock brushes with a PS block-solvent should induce a rearrangement that places PS segments at the air interface; the expected advancing contact angle for a PS-rich surface is

Table 3 Calculated δ based on group contribution methods compared to experimental values [a]

Monomer	Calc. δ $(J/cm^3)^{1/2}$	Exp. δ $(J/cm^3)^{1/2}$
Methyl acrylate (MA)	19.9	19.9–21.3
Methyl methacrylate (MMA)	19.0	18.6–26.2
Styrene (S)	19.1	17.4–19.0
N,N-Dimethylacrylamide (DMA) [b]	25.2	unknown
Acrylic acid (AA) [b]	28.7	unknown
N,N-Dimethylaminoethyl methacrylate (DMAEMA) [b]	19.5	unknown
Hydroxyethyl methacrylate (HEMA) [b]	24.8	unknown
Heptadecafluorodecyl acrylate (HFA)	14.1	unknown
Pentafluoropropyl acrylate (PFA)	16.5	unknown
Trifluorethyl acrylate (TFA)	17.5	unknown
Pentafluorostyrene (PFS)	unknown	16.77 [c]

[a] All values obtained from published work by Van Krevelen or calculated using values of Hoftyzer and Van Krevelen [66]
[b] Values calculated by the method of Fedors [66]
[c] Value obtained from work by Su [67]

Table 4 Solvent induced rearrangement of diblock brushes

Block copolymer brush [a,b]	Starting Θ_a	Θ_a After cyclohexane treatment
Si/SiO$_2$//PS-b-PMMA [46] (26) (9)	75	98
Si/SiO$_2$//PS-b-PMA [47] (24) (9)	68	98
Si/SiO$_2$//PS-b-PDMAEMA [47] (27) (3)	63	98
Si/SiO$_2$//PS-b-PDMA [51] (18) (12)	42	65
Si/SiO$_2$//PS-b-PPFA [53] (13) (5)	112	97
Si/SiO$_2$//PS-b-PHFA [53] (10) (6)	127	110
Si/SiO$_2$//PS-b-PTFA [53] (17) (6)	101	97
Si/SiO$_2$//PS-b-PPFS [53] (16) (5)	121	100
Si/SiO$_2$//PS-b-PAA [52] (21) (8)	24	48 [c]

[a] See Table 3 for abbreviations
[b] Numbers in parentheses correspond to the individual film thickness for each block, given in nm
[c] Anisole was used as the PS-selective solvent for this system

100°. Several systems do not display this behavior (indicating little rearrangement of the diblock brush) and these correspond to blocks where $\Delta\delta > 2.6$ relative to PS. The data in Table 4 provide a better understanding of the relationship between rearrangement of diblock brushes and the relative miscibility of the two blocks. The solubility parameter is a useful predictor of brush rearrangement; the greater the difference in δ between two blocks, the less rearrangement. We have also synthesized several diblock brushes where the first block is poly(methyl acrylate) (PMA) (Table 1), but the total number of diblock systems is considerably smaller and we have excluded these systems for this discussion.

4.3
Dynamics of Surface Reorganization

To better understand the time-scale of diblock brush reorganization, we prepared semifluorinated diblock copolymer brushes where the outer block (at the air interface) was a semifluorinated block composed of poly(pentafluorostyrene) (PPFS), poly(heptadecafluorodecyl acrylate) (PHFA), poly(penta-fluoropropyl acrylate) (PPFA), or poly(trifluoroethyl acrylate) (PTFA) [53]. The block at the silicate interface was either PS or PMA. Treatment of the diblock systems with block-selective solvents produced predictable changes in water contact angles except for those diblock brushes based on PHFA. All of these systems were fully characterized by XPS, tensiometry, ellipsometry,

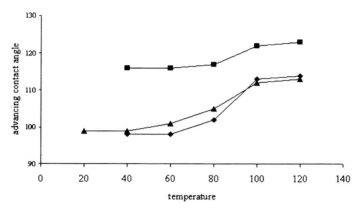

Fig. 4 Dependence of the advancing water contact angle on annealing temperature for PS-based diblock copolymer brush layers: (*filled squares*) Si/SiO$_2$//PS-*b*-PHFA, (*filled triangles*) Si/SiO$_2$//PS-*b*-PPFA, (*filled diamonds*) Si/SiO$_2$//PS-*b*-PPFS. *Lines* added as guide for the eye

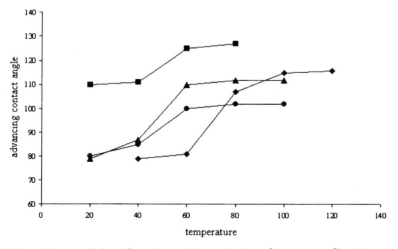

Fig. 5 Dependence of the advancing water contact angle on annealing temperature for PMA-based diblock copolymer brush layers: (*filled squares*) Si/SiO$_2$//PMA-*b*-PHFA, (*filled circles*) Si/SiO$_2$//PMA-*b*-PTFA, (*filled diamonds*) Si/SiO$_2$//PMA-*b*-PPFS, (*filled triangles*) Si/SiO$_2$//PMA-*b*-PPFA. *Lines* added as guide for the eye

AFM and ATR FTIR. Recognizing that semifluorinated blocks prefer the hydrophobic air interface, we induced a brush rearrangement of the diblock with a selective solvent for the bottom block (PS or PMA). Thermal treatment of these switched systems provided information about the time scale for equilibration to an extended state, in the absence of a solvent (see Figs. 4 and 5) [68]. Tables 5 and 6 present switching information for four out of the eight systems studied. These four diblock brushes demonstrated the greatest changes in contact angles.

Figures 4 and 5 illustrate the thermal rearrangement of semi-fluorinated diblocks after the diblock systems were treated with solvent that is selective for the lower block (PMA or PS). When either PS ($T_g = 100\,^\circ$C) or PPFS

Table 5 Solvent treatment of PS-based diblock semifluorinated brushes [a,b]

Solvent [a]	Si/SiO$_2$//PS-b-PPFS [c]		Si/SiO$_2$//PS-b-PPFA [c]	
	Θ_a	Θ_r	Θ_a	Θ_r
1st Fluorobenzene/ Trifluorotoluene [c]	121	90	112	92
1st Cyclohexane	101	85	97	78
2nd Fluorobenzene/ Trifluorotoluene [c]	119	88	113	92
2nd Cyclohexane	102	87	96	77

[a] The standard deviation of contact angles was $< 2^\circ$
[b] Sample immersed in solvent at 60 $^\circ$C for 1 h
[c] PS-b-PPFS brush was treated with fluorobenzene and all other brushes were treated with trifluorotoluene

Table 6 Solvent treatment of PMA-based semifluorinated diblock brushes

Solvent [a]	Si/SiO$_2$//PMA-b-PPFS [b]		Si/SiO$_2$//PMA-b-PPFA [b]	
	Θ_a	Θ_r	Θ_a	Θ_r
1st Fluorobenzene/ Trifluorotoluene [c]	115	88	111	94
1st Acetone/ Ethyl Acetate [d]	79	67	81	65
2nd Fluorobenzene/ Trifluorotoluene [c]	118	92	112	95
2nd Acetone/ Ethyl Acetate [d]	82	68	80	66

[a] Sample immersed in solvent at 60 $^\circ$C for 1 h
[b] The standard deviation of contact angles was $< 2^\circ$
[c] PMA-b-PPFS brush was treated with fluorobenzene and all other brushes were treated with trifluorotoluene

($T_g = 105°$) are present in the diblock brush, higher temperatures are required to effect rearrangement. Logically, the diblock brush with PMA ($T_g = 10\,°C$) and PPFA ($T_g = -26\,°C$) rearranges readily at lower temperatures. In addition to using heat as a switching stimulus, we have also effected the same remigration of a semifluorinated block to the air interface by subjecting a sample to supercritical CO_2. Fluoroacrylates are known to be highly soluble in supercritical CO_2 [69].

5
Summary

We have established a relationship between film thickness and the molecular weight of the polymer brush. We better understand how the relative film thickness of diblock copolymer brushes correlates with the dimensions of pinned micelle structures. Probably the most significant result is a better understanding of the relationship between rearrangement of diblock brushes and the relative miscibility of the two blocks. We have demonstrated that δ is a useful predictor of brush rearrangement; the greater the difference in between two blocks, the less significant the rearrangement. Lastly, we have demonstrated the other external stimuli besides block-selective solvents can be used to induce brush reorganization, namely temperature and treatment with supercritical CO_2.

Acknowledgements The authors would like to acknowledge the financial support of the National Science Foundation (DMR-0729977, DMR-0423786). We also acknowledge Professor Stephen Z. D. Cheng (The University of Akron) for AFM measurements and Dr. Wayne Jennings (Case Western Reserve University MATNET Center) for help with XPS.

References

1. Milner ST (1991) Science 251:905
2. Halpern A, Tirrell M, Lodge TP (1992) Adv Polym Sci 100:31
3. Zhao B, Brittain WJ (2000) Prog Polym Sci 25:677
4. Advincula RC, Brittain WJ, Caster KC, Rühe J (2004) Polymer brushes. Wiley, Weinheim
5. Mansky P, Liu Y, Huang E, Russell TP, Hawker CJ (1997) Science 275:1458
6. Delmarche E, Michel B, Biebuyck H, Gerber C (1996) Adv Mater 8:719
7. Dubois LH, Nuzzo RG (1992) Annu Rev Phys Chem 43:437
8. Matyjaszewski K (2000) In: Matyjaszewski K (ed) ACS symposium series 768. ACS, Washington, p 2
9. Prucker O, Rühe J (1998) Macromolecules 31:592
10. Prucker O, Rühe J (1998) Macromolecules 31:602
11. Wu T, Efimenko K, Genzer J (2002) J Am Chem Soc 124:9394

12. Desai SM, Solanky SS, Mandale AB, Rathore K, Singh RP (2003) Polymer 44:7645
13. Kaholek M, Lee WK, LaMattina B, Caster KC, Zauscher S (2004) Nano Lett 4:373
14. Kaholek M, Lee WK, Ahn SJ, Ma HW, Caster KC, LaMattina B, Zauscher S (2004) Chem Mater 16:3688
15. Kizhakkedathu JN, Norris-Jones R, Brooks DE (2004) Macromolecules 37:734
16. Farhan T, Huck WTS (2004) Eur Polym J 40:1599
17. Zhu X, De Graaf J, Winnik FM, Leckband D (2004) Langmuir 20:1459
18. Xu FJ, Zhong SP, Yung KYL, Kang ET, Neoh KG (2004) Biomacromolecules 5:2392
19. Zhang HJ, Ito Y (2001) Langmuir 17:8336
20. Ryan AJ, Crook CJ, Howse JR, Topham P, Jones RAL, Geoghegan M, Parnell AJ, Ruiz-Perez L, Martin SJ, Cadby A, Menelle A, Webster JRP, Gleeson AJ, Bras W (2005) Faraday Disc 128:55
21. Zhao B, Haasch RT, MacLaren S (2004) J Am Chem Soc 126:6124
22. Zhao B, He T (2003) Macromolecules 36:8599
23. Ionov L, Minko S, Stamm M, Gohy JF, Jerome R, Scholl A (2003) J Am Chem Soc 125:8302
24. Ionov L, Stamm M, Minko S, Hoffmann F, Wolff T (2004) Macromol Symp 210:229
25. Draper J, Luzinov I, Minko S, Tokarev I, Stamm M (2004) Langmuir 20:4064
26. Motornov M, Minko S, Eichhorn KJ, Nitschke M, Simon F, Stamm M (2003) Langmuir 19:8077
27. Luzinov I, Minko S, Tsukruk VV (2004) Prog Polym Sci 29:635
28. Stamm M, Minko S, Tokarev I, Fahmi A, Usov D (2004) Macromol Symp 214:73
29. Zhulina EB, Singh C, Balazs AC (1996) Macromolecules 29:6338
30. Zhulina EB, Singh C, Balazs AC (1996) Macromolecules 29:8904
31. Zhao B, Brittain WJ, Zhou W, Cheng SZD (2000) Macromolecules 33:8821
32. Zhao B, Brittain WJ, Zhou W, Cheng SZD (2000) J Am Chem Soc 122:240
33. Boyes SG, Brittain WJ, Weng X, Cheng SZD (2002) Macromolecules 35:4960
34. Huang W, Kim J-B, Baker GL, Bruening ML (2003) Nanotechnology 14:1075
35. Husseman M, Malmstrom EE, McNamara M, Mate M, Mecerreyes O, Benoit DG, Hedrick JL, Mansky P, Huang E, Russell TP, Hawker CJ (1999) Macromolecules 32:1424
36. Matyjaszewski K, Miller PJ, Shukula N, Immaraporn B, Gelman A, Luokala BB, Siclovan TM, Kickelbick G, Vllant T, Hoffmann H, Pakula T (1999) Macromolecules 32:8716
37. Zhao H, Farrell BP, Shipp DA (2004) Polymer 45:4473
38. Yu WH, Kang ET, Neoh KG, Zhu S (2003) J Phys Chem B 107:10198
39. Kong X, Kawai T, Abe J, Iyoda T (2001) Macromolecules 34:1837
40. Tomlinson MR, Genzer J (2003) Chem Comm 1350
41. Huang W, Kim J-B, Baker GL, Bruening ML (2002) Macromolecules 35:1175
42. Prokhoroava SA, Kopyshev A, Ramakrishnan A, Zhang H, Rühe J (2003) Nanotechnology 14:1098
43. Osborne VL, Jones DM, Huck WTS (2002) Chem Comm 1838
44. Quirk RP, Mathers RT, Cregger T, Foster MD (2002) Macromolecules 35:9964
45. Advincula R, Zhou Q, Park M, Wang S, Mays J, Sakellariou G, Pipas S, Hadjichristidis N (2002) Langmuir 18:8672
46. Zhao B, Brittain WJ (1999) J Am Chem Soc 121:3557
47. Zhao B, Brittain WJ (2000) Macromolecules 33:8813
48. Sedjo RA, Mirous BK, Brittain WJ (2000) Macromolecules 33:1492
49. Moineau G, Dubois P, Jerome R, Senninger T, Teyssie P (1998) Macromolecules 31:545
50. Xia J, Matyjaszewski K (1997) Macromolecules 30:7692

51. Baum M, Brittain WJ (2002) Macromolecules 35:610
52. Boyes SG, Akgun B, Brittain WJ, Foster MD (2003) Macromolecules 36:9539
53. Granville AM, Boyes SG, Akgun B, Foster MD, Brittain WJ (2004) Macromolecules 37:2790
54. Tsujii T, Ejaz M, Sato K, Goto A, Fukuda T (2001) Macromolecules 34:8872
55. Baum M (2002) PhD thesis, University of Akron
56. von Werne T, Patten TE (2001) J Am Chem Soc 123:7497
57. von Werne T, Patten TE (1999) J Am Chem Soc 121:7409
58. Schmidt R, Zhao T, Green J-B, Dyer DJ (2002) Langmuir 18:1281
59. Vaughn J, Boyes SG, Brittain WJ (2003) unpublished work
60. Ejaz M, Tsujii Y, Fukuda T (2001) Polymer 42:6811
61. Kim J-B, Bruening ML, Baker GL (2000) J Am Chem Soc 122:7616
62. Dong H, Marko JF, Witten TA (1994) Macromolecules 27:6428
63. Zhulina EB, Balazs AC (1996) Macromolecules 29:2667
64. Zhulina EB, Singh C, Balazs AC (1996) Macromolecules 29:8254
65. Gersappe D, Fasolka M, Israels R, Balazs AC (1995) Macromolecules 28:4753
66. Van Krevlen DW, Hoftyzer PJ (1976) Properties of polymers, chap 4 and 7. Elsevier, New York
67. Su W (1991) PhD thesis, The University of Akron
68. Granville AM, Boyes SG, Akgun B, Foster MD, Brittain WJ (2005) Macromolecules 38:3263
69. Behles JA, Desimone JM (2001) Pure Appl Chem 73:1281

Adv Polym Sci (2006) 198: 149–183
DOI 10.1007/12_062
© Springer-Verlag Berlin Heidelberg 2005
Published online: 20 December 2005

Theoretical Approaches to Neutral and Charged Polymer Brushes

Ali Naji[1] · Christian Seidel[2] · Roland R. Netz[1] (✉)

[1] Dept. of Physics, Technical University of Munich, James Franck Str, 85478 Garching, Germany
netz@ph.tum.de

[2] Max-Planck-Institute of Colloids and Interfaces, 14424 Potsdam, Germany

Abstract Neutral or charged polymers that are densely end-grafted to surfaces form brush-like structures and are highly stretched under good-solvent conditions. We discuss and compare relevant results from scaling models, self-consistent field methods and MD simulations and concentrate on the conceptual simple case of planar substrates. For neutral polymers the main quantity of interest is the brush height and the polymer density profile, which can be well predicted from self-consistent calculations and simulations. Charged polymers (polyelectrolytes) are of practical importance since they are soluble in water. Counterion degrees of freedom determine the brush behavior in a decisive way and lead to a strong and nonlinear swelling of the brush.

Keywords Brushes · Polyelectrolytes · Scaling theory · Self-consistent field theory · Simulation techniques

Abbreviations

a	Kuhn length or effective monomer size
d	height of counterion layer
f	fractional charge of the chain $0 < f < 1$
F	free energy in units of $k_B T$ (per chain or unit area)
h	height of brush
$k_B T$	thermal energy
L	contour length of a chain
$\ell_B = e^2/(4\pi\varepsilon k_B T)$	
N	polymerization index
R	end-to-end polymer chain radius
R_0	end-to-end radius of an ideal polymer
R_F	Flory radius of a self-avoiding chain
v_2	2nd virial coefficient of monomers in solution
κ^{-1}	Debye–Hückel screening length
ν	Flory exponent for the polymer size
Π	osmotic pressure, rescaled by $k_B T$
ρ_a	grafting density of a polymer brush
$\rho(z)$	monomer density at distance z from grafting surface
σ	Lennard-Jones diameter in simulation

1
Introduction

Polymers are long, chain-like molecules that consist of repeating subunits, the so-called monomers [1–4]. In many situations, all monomers of a polymer are alike, showing for example the same tendency to adsorb to a substrate [5]. For industrial applications, one is often interested in *end-functionalized polymers* that are attached with one end only to the substrate [6, 7]. Industrial interest comes from the need to stabilize particles and surfaces against flocculation. In end-grafted polymer structures the stabilization power is greatly enhanced as compared with adsorbed layers of polymers, where each monomer is equally attracted to the substrate. The main reason is that bridging of polymers between two approaching surfaces and creation of polymer loops on the same surface is very frequent in the case of polymer adsorption and eventually leads to attraction between two particle surfaces and thus destabilization. This does not occur if the polymer is grafted by its end to the surface and the monomers are chosen such that they do not particularly adsorb to the surface. In biology, brush-like polymer structures are encountered as the surface coating of endothelial cells and regulate the adsorption and migration of various particles and bio-molecules from the blood stream to the vascular cells.

Experimentally, two basic ways of building a grafted polymer layer can be distinguished: In the first, the polymerization is started from the surface with some suitably chosen surface-linked initiator. The advantage of this *grafting-*

from procedure is that only monomers have to diffuse through the forming brush layer and thus the reaction kinetics are fast. In the second route one attaches polymers with special end-groups that act as anchors on the surface. This *grafting-to* procedure is subject to slow kinetics during the formation stage since whole polymers have to diffuse through the natant grafting layer, but benefits from a somewhat better control over the brush constitution and chemical composition. One distinguishes *physical* adsorption of end-groups that favor the substrate, for example zwitter-ionic end-groups attached to polystyrene chains that lead to binding to mica in organic solvents such as toluene or xylene [8]. A stronger and thus more stable attachment is possible with *covalently* end-grafted chains, for example poly-dimethylsiloxane chains which carry hydroxyl end groups and undergo condensation reactions with silanols of a silica surface [9]. One can also employ a suitably chosen diblock copolymer where one block adsorbs on the substrate while the other is repelled from it [10]. An example is furnished by polystyrene-poly(vinyl-pyridine) (PS-PVP) diblocks in the selective solvent toluene, which is a bad solvent for the PVP block and promotes strong adsorption on a quartz substrate, but acts as a good solvent for the PS block and disfavors its adsorption on the substrate [11]. In a slight modification one uses diblock copolymers that are anchored at the liquid–air [12, 13] or at a liquid–liquid interface of two immiscible liquids [14]. This scenario offers the advantage that the grafting density can (for the case of strongly anchored polymers) be varied by lateral compression (like a Langmuir mono-layer) and that the lateral surface pressure can be directly measured. The lateral pressure is an important thermodynamic quantity and allows detailed comparison with theoretical predictions. A well studied example is that of a diblock copolymer of polystyrene—polyethylene oxide (PS-PEO) [13]. The PS block is shorter and functions as an anchor at the air–water interface because it is immiscible in water. The PEO block is miscible in water but because of attractive interaction with the air–water interface it forms a quasi-two dimensional layer at very low surface coverage. As the surface pressure increases and the area per polymer decreases, the PEO block is expelled from the surface and forms a polymer brush.

In this theoretical chapter we simplify the discussion by assuming that the polymers are irreversibly grafted at one of their chain ends to the substrate, we only mention in passing a few papers on the kinetics of grafting. The substrate is assumed to be solid, planar and impenetrable to the polymer monomers, we briefly cite some results obtained for curved substrates. We limit the discussion to good solvent conditions and neglect any attractive interactions between the polymer chains and the surface. Charged polymers are interesting from the application point of view, since they allow for water-based formulations of organic substances which are advantageous for economical and ecological reasons. Recent years have seen a tremendous research activity on charged polymers in bulk [15–18] and at interfaces [5, 19]. We therefore treat neutral brushes as well as charged ones.

The characteristic parameter for brush systems is the anchoring or graft-ing density ρ_a, which is the inverse of the average area available for each polymer at the surface. For small grafting densities, $\rho_a < \rho_a^*$, the polymer chains will be far apart from each other and hardly interact, as schematically shown in Fig. 1a. The polymers in this case form well-separated *mushrooms* at the surface. The grafting density at which chains just start to overlap is determined by $\rho_a^* \sim R^{-2}$ where R is the typical radius or size of a chain. In good solvent conditions (that is for swollen chains), the chain radius fol-lows the Flory prediction $R_F \sim aN^{3/5}$ where N is the polymerization index or monomer number of the chain, and a is a characteristic microscopic length scale of the polymer that incorporates monomer size as well as backbone stiffness. The crossover grafting density for a polymer under good-solvent conditions follows as $\rho_a^* \sim a^{-2}N^{-6/5}$. For large grafting densities, $\rho > \rho^*$, the chains are strongly overlapping. This situation is depicted in Fig. 1b. Since we assume the solvent to be good, monomers repel each other. The lateral separation between the polymer coils is fixed by the grafting density, so that the polymers extend away from the grafting surface in order to avoid each other. The resulting structure is called a *polymer brush*, with a vertical height h which greatly exceeds the unperturbed coil radius R [20–22]. Similar stretched structures occur in many other situations, such as diblock copoly-mer melts in the strong segregation regime [6, 23], or star polymers under good solvent conditions [24]. Theory is mostly concerned with predicting the layer height h, but also the detailed monomer density profile and resulting

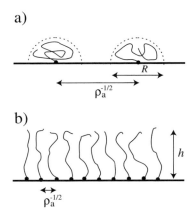

Fig. 1 For grafted chains, one distinguishes between: **a** the mushroom regime, where the distance between chains, $\rho_a^{-1/2}$, is larger than the size R of a polymer coil; and, **b** the brush regime, where the distance between chains is smaller than the unperturbed coil size. Here, the chains are stretched away from the surface due to repulsive interactions between monomers. The brush height h scales linearly with the polymerization index, $h \sim N$, and thus is larger than the unperturbed coil radius R which in the good solvent regime scales according to Flory as $R_F \sim aN^{\nu}$ with $\nu = 3/5$

forces of lateral or vertical brush compression as a function of the various system parameters.

The understanding of grafted polymer systems progressed substantially with the advent of experimental techniques such as: surface force balance [8], small angle neutron scattering [9], neutron [11, 25] and X-ray [26] diffraction, and ellipsometry [14]. Of equal merit was the advancement in the theoretical methodology ranging from field theoretical methods and scaling arguments to numerical simulations, which will be amply reviewed in this chapter.

2
Polymer Basics

The main parameters used to describe a polymer chain are the polymerization index N, which counts the number of repeat units or monomers along the chain, and the size of one monomer or the distance between two neighboring monomers. The monomer size ranges from a few Angstroms for synthetic polymers to a few nanometers for biopolymers. The simplest theoretical description of flexible chain conformations is achieved with the so-called freely-jointed chain (FJC) model, where a polymer consisting of $N + 1$ monomers is represented by N bonds defined by bond vectors r_j with $j = 1, ...N$. Each bond vector has a fixed length $|r_j| = a$ corresponding to the Kuhn length, but otherwise is allowed to rotate freely and independently of its neighbors, as is schematically shown in Fig. 2 (top). This model of course only gives a coarse-grained description of real polymer chains, but we will later see that by a careful interpretation of the Kuhn length a and the monomer number N, an accurate description of the large-scale properties of real polymer chains is possible. The main advantage is that due to the simplicity of the FJC model, many interesting observables (such as chain size or distribution functions) can be calculated with relative ease. We demonstrate this by calculating the mean end-to-end radius of such a FJC polymer. Fixing one of the chain ends at the origin, the position of the $(k + 1)$-th monomer is given by the vectorial sum

$$R_k = \sum_{j=1}^{k} r_j. \tag{1}$$

Because two arbitrary bond vectors are uncorrelated in this simple model, the thermal average over the scalar product of two different bond vectors vanishes, $\langle r_j \cdot r_k \rangle = 0$ for $j \neq k$, while the mean squared bond vector length is simply given by $\langle r_j^2 \rangle = a^2$. It follows that the mean squared end-to-end radius R_N^2 is proportional to the number of monomers,

$$R_0^2 \equiv \langle R_N^2 \rangle = Na^2 = La, \tag{2}$$

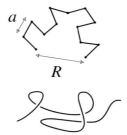

Fig. 2 *Top*: Freely jointed chain (FJC) model, where N bonds of length a are connected to form a flexible chain with a certain end-to-end distance R. *Bottom*: In the simplified model, appropriate for more advanced theoretical calculations, a continuous line is governed by some bending rigidity or line tension. This continuous model can be used when the relevant length scales are much larger than the monomer size

where the contour length of the chain is given by $L = Na$. R_0 denotes the mean end-to-end radius of an ideal chain, and according to Eq. 2, it scales as $R_0 = aN^{1/2}$. Experimentally one often knows, via knowledge of the chemical structure and polymer mass, the length L and, via light scattering, the radius R of a chain. Using the above scaling results, valid for a noninteracting chain, the Kuhn length follows as $a = R_0^2/L$ and the effective monomer number as $N = L/a$, which allows treatment of real chains with a complicated local conformational structure within the FJC model. Note that the so-determined Kuhn length a is usually larger than the actual monomer-size, since it takes back-bone stiffness effects into account. Likewise, the determined effective monomer number N is typically smaller than the actual (chemical) number of monomers in the chain.

In many theoretical calculations aimed at elucidating large-scale properties, the simplification is carried even a step further and a continuous model is used, as schematically shown in Fig. 2 (bottom). In such models the polymer backbone is replaced by a continuous line and all microscopic details are neglected. The chain is then only characterized by its length L and radius R.

2.1
Polymer Swelling and Collapse

The models discussed so far describe ideal chains and do not account for interactions between monomers which typically consist of some short-ranged repulsion and long-ranged attraction. Including these interactions will give a different scaling behavior for long polymer chains. The end-to-end radius, $R = \sqrt{\langle R_N^2 \rangle}$, can be written for $N \gg 1$ as

$$R \simeq aN^\nu, \tag{3}$$

which defines the so-called *swelling exponent* ν. As we have seen, for an ideal polymer chain (no interactions between monomers), Eq. 2 implies $\nu = 1/2$. This situation is realized for experimental polymers at a certain temperature or solvent conditions when the attraction between monomers exactly cancels the steric repulsion (which is due to the fact that the monomers cannot penetrate each other). This situation can be achieved in the condition of *theta solvents*. In *good solvents*, on the other hand, the monomer-solvent interaction is more attractive than the monomer-monomer interaction, in other words, the monomers try to avoid each other in solution. As a consequence, single polymer chains in good solvents have swollen spatial configurations dominated by the steric repulsion, characterized by an exponent $\nu \simeq 3/5$, leading to the Flory radius $R_F \sim aN^{3/5}$. This spatial size of a polymer coil is much smaller than the extended contour length $L = aN$ but larger than the size of an ideal chain $R_0 = aN^{1/2}$, i.e. $R_0 \ll R_F \ll L$ for large N. The reason for this behavior is conformational entropy (which prevents full stretching of the chain) combined with the favorable interaction between monomers and solvent molecules in good solvents (which leads to a more open structure than an ideal chain). In the opposite case of *bad or poor solvent* conditions, the effective interaction between monomers is attractive, leading to collapse of the chains and to their precipitation from solution. In this case, the polymer volume, like any space filling object embedded in three-dimensional space, scales proportional to its weight as $R^3 \sim N$, yielding an exponent $\nu = 1/3$.

The standard way of taking into account interactions between monomers is the Flory theory, which treats these interactions on an approximate mean-field level [1–3]. Let us first consider the case of repulsive interactions between monomers, which can be described by a positive second-virial coefficient v_2 and corresponds to the good solvent condition. For pure hard-core interactions and with no additional attractions between monomers, the second virial coefficient (which corresponds to the excluded volume) is of the order of the monomer volume, i.e. $v_2 \sim a^3$. The repulsive interaction between monomers, which tends to swell the chain, is counteracted and balanced by the ideal chain elasticity, which is brought about by the entropy loss associated with stretching the chain. The origin is that the number of polymer configurations having an end-to-end radius of the order of the unperturbed end-to-end radius is large. These configurations are entropically favored over configurations characterized by a large end-to-end radius, for which the number of possible polymer conformations is drastically reduced. The standard Flory theory for a flexible chain of radius R is based on writing the free energy F of a chain (in units of the thermal energy $k_B T$) as a sum of two terms (omitting numerical prefactors)

$$F \simeq \frac{R^2}{R_0^2} + v_2 R^3 \left(\frac{N}{R^3} \right)^2 , \qquad (4)$$

where the first term is the entropic elastic energy associated with swelling a polymer chain to a radius R, proportional to the effective spring constant of an ideal chain, $k_B T/R_0^2$, and the second term is the second-virial repulsive energy proportional to the coefficient v_2, and the segment density squared. It is integrated over the volume R^3. The optimal radius R is calculated by minimizing this free energy and gives the swollen radius

$$R_F \sim a(v_2/a^3)^{1/5} N^\nu ,\qquad (5)$$

with $\nu = 3/5$. For purely steric interactions with $v_2 \simeq a^3$ we obtain $R \sim aN^\nu$. For weakly interacting monomers, $0 < v_2 < a^3$, one finds that the swollen radius Eq. 5 is only realized above a minimal monomer number $N \simeq (v_2/a^3)^{-2}$ below which the chain statistics is unperturbed by the interaction and the scaling of the chain radius is Gaussian and given by Eq. 2.

In the opposite limit of a negative second virial coefficient, $v_2 < 0$, corresponding to the bad or poor solvent regime, the polymer coil will be collapsed due to attraction between monomers. In this case, the attraction term in the free energy is balanced by the third-virial term in a low-density expansion (where we assume that $v_3 > 0$),

$$F \simeq v_2 R^3 \left(\frac{N}{R^3}\right)^2 + v_3 R^3 \left(\frac{N}{R^3}\right)^3 .\qquad (6)$$

Minimizing this free energy with respect to the chain radius one obtains

$$R \simeq (v_3/|v_2|)^{1/3} N^\nu \qquad (7)$$

with $\nu = 1/3$. This indicates the formation of a compact globule, since the monomer density inside the globule, $\rho \sim N/R^3$, is independent of the chain length. The minimal chain length to observe a collapse behavior is $N \sim (v_3/a^3 v_2)^2$. For not too long chains and a second virial coefficient not too much differing from zero, the interaction is irrelevant and one obtains effective Gaussian or ideal behavior. It should be noted, however, that even small deviations from the exact theta conditions (defined by strictly $v_2 = 0$) will lead to chain collapse or swelling for very long chains.

2.2
Charged Polymers

A polyelectrolyte (PE) is a polymer with a fraction f of charged monomers. When this fraction is small, $f \ll 1$, the PE is weakly charged, whereas when f is close to unity, the polyelectrolyte is strongly charged. There are two common ways to control f [17]. One way is to polymerize a heteropolymer mixing strongly acidic and neutral monomers as building blocks. Upon contact with water, the acidic groups dissociate into positively charged protons (H^+) that bind immediately to water molecules, and negatively charged monomers.

Although this process effectively charges the polymer molecules, the counterions make the PE solution electro-neutral on larger length scales. The charge distribution along the chain is quenched during the polymerization stage, and it is characterized by the fraction of charged monomers on the chain, f. In the second way, the PE is a weak polyacid or polybase. The effective charge of each monomer is controlled by the pH and the salt concentration of the solution [27]. Moreover, this annealed fraction depends on the local electric potential which is in particular important for adsorption or binding processes since the local electric potential close to a strongly charged surface [28] or a second charged polymer can be very different from its value in the bulk solution and therefore modify the polyelectrolyte charge [29].

Counterions are attracted to the charged polymers via long-ranged Coulomb interactions; this physical association typically leads to a rather loosely bound counterion cloud around the PE chain. Because of this background of a polarizable and diffusive counterion cloud, there is a strong influence of the counterion distribution on the PE structure and vice versa. Counterions contribute significantly towards bulk properties, such as the osmotic pressure, and their translational entropy is responsible for the generally good water solubility of charged polymers. In addition, the statistics of PE chain conformations is governed by intra-chain Coulombic repulsion between charged monomers; this results in a more extended and swollen conformation of PE's as compared to neutral polymers and gives rise to the characteristically high viscosity of polyelectrolyte solutions (hence their use as viscosifiers in the food industry).

For polyelectrolytes, electrostatic interactions provide the driving force for their salient features and have to be included in any theoretical description. The reduced electrostatic interaction between two point-like charges can be written as $q_1 q_2 v(r)$ where

$$v(r) = \ell_B / r \tag{8}$$

is the Coulomb interaction between two elementary charges in units of $k_B T$ and q_1 and q_2 are the valencies (or the reduced charges in units of the elementary charge e). The Bjerrum length ℓ_B is defined as

$$\ell_B = \frac{e^2}{4\pi\varepsilon k_B T}, \tag{9}$$

where ε is the medium dielectric constant. It denotes the distance at which the Coulombic interaction between two unit charges in a dielectric medium is equal to thermal energy ($k_B T$). It is a measure of the distance below which the Coulomb energy is strong enough to compete with the thermal fluctuations; in water at room temperatures, one finds $\ell_B \approx 0.7$ nm.

In biological systems and most industrial applications, the aqueous solution contains in addition to the counterions mobile salt ions. Salt ions of opposite charge are drawn to the charged object and modify the counterion cloud around it. They effectively reduce or *screen* the charge of the object. The effective (screened) electrostatic interaction between two charges q_1 and q_2 in the presence of salt ions and a polarizable solvent can be written as $q_1 q_2 v_{DH}(r)$, with the Debye–Hückel (DH) potential $v_{DH}(r)$ given (in units of $k_B T$) by

$$v_{DH}(r) = \frac{\ell_B}{r} e^{-\kappa r}. \tag{10}$$

The exponential decay is characterized by the screening length κ^{-1}, which is related to the salt concentration c_{salt} by $\kappa^2 = 8\pi q^2 \ell_B c_{salt}$, where q denotes the valency of $q : q$ salt. At physiological conditions the salt concentration is $c_{salt} \approx 0.1$ M and for monovalent ions ($q = 1$) this leads to $\kappa^{-1} \approx 1$ nm. This means that although the Coulombic interactions are long-ranged, in physiological conditions they are highly screened above length scales of a few nanometers, which results from multi-body correlations between ions in a salt solution.

For charged polymers, the effective bending stiffness and thus the Kuhn length is increased due to electrostatic repulsion between monomers [30–36]. This effect modifies considerably not only the PE behavior in solution but also their adsorption characteristics [37].

A peculiar phenomenon occurs for highly charged PE's and is known as the Manning condensation of counterions [38–40]. This phenomenon constitutes a true phase transition in the absence of added salt ions [41]. For a single rigid PE chain represented by an infinitely long and straight cylinder with a linear charge density τ larger than the threshold

$$\ell_B \tau q = 1, \tag{11}$$

where q is the counterion valency, it was shown that counterions condense on the oppositely charged cylinder even in the limit of infinite system size. For a solution of stiff charged polymers this corresponds to the limit where the inter-chain distance tends to infinity. This effect is not captured by the linear Debye–Hückel theory. A simple heuristic way to incorporate the non-linear Manning condensation is to replace the bare linear charge density τ by the renormalized one, $\tau_{renorm} = 1/(q\ell_B)$, whenever $\ell_B \tau q > 1$ holds. This procedure, however, is not totally satisfactory at high salt concentrations [42, 43]. Also, real polymers have a finite length, and are neither completely straight nor in the infinite dilution limit [44]. Still, Manning condensation has an experimental significance for polymer solutions [45] because thermodynamic quantities, such as counterion activities [46] and osmotic coefficients [47], show a pronounced signature of Manning condensation. Locally, polymer segments can be considered as straight over length scales comparable to the persistence length. The Manning condition Eq. 11 usually denotes a region where the binding of counterions to charged chain sections begins to deplete the solution from free counterions.

3
Neutral Grafted Polymers

3.1
Scaling Approach

The scaling behavior of the brush height h can be analyzed using a Flory-like mean-field theory, which is a simplified version of the original Alexander theory [21] for polymer brushes. The stretching of the chain leads to an entropic free energy loss of h^2/R_0^2 per chain, and the repulsive energy density due to unfavorable monomer-monomer contacts is proportional to the squared monomer density times the excluded-volume parameter v_2. The derivation is thus analogous to the calculation of the Flory radius of a chain in good solvent shown in Sect. 2.1, except that now the grafting density ρ_a plays a decisive role and controls the amount of stretching of the chains. The free energy per chain (and in units of $k_B T$) is then

$$F \simeq \frac{h^2}{a^2 N} + v_2 \left(\frac{\rho_a N}{h}\right)^2 \frac{h}{\rho_a} . \tag{12}$$

The equilibrium height is obtained by minimizing Eq. 12 with respect to h, and the result is

$$h_0 = N \left(2 v_2 a^2 \rho_a / 3\right)^{1/3} , \tag{13}$$

where the numerical constants have been added for numerical convenience in the following considerations. The height of the brush scales linearly with the polymerization index N, a clear signature of the strong stretching of the polymer chains, as was originally obtained by Alexander [21]. At the overlap threshold, $\rho_a^* \sim a^{-2} N^{-6/5}$, the height scales as $h_0 \sim N^{3/5}$, and thus agrees with the scaling of an unperturbed chain radius in a good solvent, Eq. 5, as it should. The simple scaling calculation predicts the brush height h correctly in the asymptotic limit of long chains and strong overlap. It has been confirmed by experiments [8, 9, 11] and computer simulations [48, 49].

3.2
Mean-Field Calculation

The above scaling result assumes that all chains are stretched to exactly the same height, leading to a step-like shape of the density profile. Monte Carlo and numerical mean-field calculations confirm the general scaling of the brush height, but exhibit a more rounded monomer density profile which goes continuously to zero at the outer perimeter [48]. A big step towards a better understanding of stretched polymer systems was made by Semenov [23], who recognized the importance of *classical paths* for such systems.

The classical polymer path is defined as the path which minimizes the free energy, for given start and end positions, and thus corresponds to the most likely path a polymer can take. The name follows from the analogy with quantum mechanics, where the classical motion of a particle is given by the quantum path with maximal probability. Since for strongly stretched polymers the fluctuations around the classical path are weak, it is expected that a theory that takes into account only classical paths, is a good approximation in the strong-stretching limit. To quantify the stretching of the brush, let us introduce the (dimensionless) interaction parameter β, defined as

$$\beta \equiv N \left(\frac{3 v_2^2 \rho_a^2}{2 a^2} \right)^{1/3} = \frac{3}{2} \left(\frac{h_0}{a N^{1/2}} \right)^2 , \qquad (14)$$

where h_0 is the brush height according to Alexander's theory, compare Eq. 13. The parameter β is proportional to the square of the ratio of the Alexander prediction for the brush height, h_0, and the unperturbed Gaussian chain radius $R_0 \sim a N^{1/2}$, and, therefore, is a measure of the stretching of the brush based on the scaling prediction. We will later see that the actual stretching of the brush is not correctly described by the scaling result for short or weakly interacting chains. Constructing a classical theory in the infinite-stretching limit, defined as the limit $\beta \to \infty$, it was shown independently by Milner et al. [50] and Skvortsov et al. [51] that the resulting monomer density profile $\rho(z)$ depends only on the vertical distance z from the grafting surface and has in fact a *parabolic* profile. Normalized to unity, the density profile is given by [50, 51]

$$\frac{\rho(z) h_0}{\rho_a N} = \left(\frac{3\pi}{4} \right)^{2/3} - \left(\frac{\pi z}{2 h_0} \right)^2 . \qquad (15)$$

The brush height, i.e., the value of z for which the monomer density becomes zero, is given by $z^* = (6/\pi^2)^{1/3} h_0$ and is thus proportional to the scaling prediction for the brush height, Eq. 13. The parabolic brush profile has subsequently been confirmed in computer simulations [48, 49] and experiments [9] as the limiting density profile in the strong-stretching limit, and constitutes one of the cornerstones in this field. Intimately connected with the density profile is the distribution of *polymer end points*, which is nonzero everywhere inside the brush (as we will demonstrate later), in contrast with the original scaling description leading to Eq. 13.

However, deviations from the parabolic profile become progressively important as the length of the polymers N or the grafting density ρ_a decreases. In a systematic derivation of the mean-field theory for Gaussian brushes [52] it was shown that the mean-field theory is characterized by a single parameter, namely the stretching parameter β. In the limit $\beta \to \infty$, the difference between the classical approximation and the mean-field theory vanishes, and one obtains the parabolic density profile. For finite β the full mean-field theory and the classical approximation lead to different results and both show deviations from the parabolic profile.

In Fig. 3 we show the density profiles (normalized to unity) for four different values of β, obtained with the full mean-field theory [52]. In (a) the distance from the grafting surface is rescaled by the scaling prediction for the brush height, h_0; and in (b) it is rescaled by the unperturbed polymer radius R_0.

For comparison, we also show the asymptotic result according to Eq. 15 as dashed lines. The self-consistent mean-field equations are solved in the continuum limit, where the results depend only on the single parameter β and direct comparison with other continuum theories becomes possible. As β increases, the density profiles approach the parabolic profile and already for $\beta = 100$ the density profile obtained within the mean-field theory is almost indistinguishable from the parabolic profile denoted by a thick dashed line in Fig. 3a. What is interesting to see is that for small values of the interaction parameter β, the numerically determined density profiles exhibit a larger brush height than the asymptotic prediction. This has to do with the fact that entropic effects, due to steric polymer repulsion from the grafting surface, are not accounted for in the infinite-stretching approximation. Experimentally, the achievable β values are below $\beta \simeq 50$, which means that deviations from the asymptotic parabolic profile are important. For moderately large values of $\beta > 10$, the classical approximation (not shown here), derived from the mean-field theory by taking into account only one polymer path per endpoint position, is still a good approximation, as judged by comparing density profiles obtained from both theories [52], except very close to the surface. Unlike mean-field theory, the classical theory misses completely the depletion effects at the substrate. Depletion effects at the substrate lead to a pronounced density depression close to the grafting surface, as is clearly visible in Fig. 3.

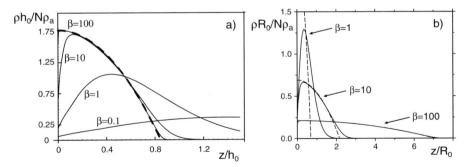

Fig. 3 Self-consistent mean-field results for the density profile (normalized to unity) of a brush for different values of the interaction parameter β. In **a** the distance from the grafting surface is rescaled by the scaling prediction for the brush height, h_0, and in **b** it is rescaled by the unperturbed polymer radius R_0. As β increases, the density profiles approach the parabolic profile (shown as *dashed lines*)

Let us now turn to the thermodynamic behavior of a polymer brush. Using the Alexander scaling model, we can calculate the free energy per chain by putting the result for the optimal brush height, Eq. 13, into the free-energy expression, Eq. 12, and obtain

$$F \sim \beta \sim N \left(v_2 \rho_a / a\right)^{2/3} . \tag{16}$$

In Fig. 4a we show the rescaled free energy per chain within the full mean-field framework (solid line) in comparison with the infinite stretching result (dotted horizontal line) and including the leading correction due to the finite entropy of the end-point distribution (broken line) [52]. In the infinite-stretching limit, i.e. for $1/\beta \to 0$, all curves converge. The free energy is not directly measurable, but the lateral pressure can be determined using the Langmuir film balance technique. The osmotic surface pressure Π, rescaled by $k_B T$, is related to F, the free energy per chain, by

$$\Pi = -\frac{\partial(F \rho_a A)}{\partial A} = \rho_a^2 \frac{\partial F}{\partial \rho_a} = \frac{2 \rho_a}{3} \left(F - \frac{\partial F / \beta}{\partial \beta^{-1}}\right) . \tag{17}$$

In Fig. 4b we show the rescaled osmotic pressure obtained within the mean-field approach (solid line). In the infinite-stretching limit, one expects a scaling as $\Pi \sim \rho_a^{5/3}$ which is shown as a dotted line. One notes that the more accurate mean-field result gives a pressure which is strictly larger than the asymptotic infinite stretching result, in agreement with previous calculations and experiments [12, 53–56]. In the presence of excluded-volume correlations, i.e., when the chain overlap is rather moderate, the scaling of the brush height h, Eq. 13, is still correctly predicted by the Alexander calculation, but the prediction for the free energy, Eq. 16, is in error. Including correlations [21], the free energy is predicted to scale as $F \sim N \rho_a^{5/6}$, leading to a pressure which scales as $\Pi \sim \rho_a^{11/6}$ in the presence of correlations. How-

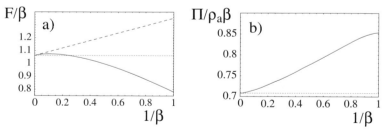

Fig. 4 a Mean-field result (*solid line*) for the rescaled brush free energy per polymer as a function of the inverse interaction parameter $1/\beta$. The infinite stretching result is indicated by a horizontal *dotted line*, the *broken straight line* denotes the infinite stretching result with the leading correction due to the finite end-point distribution entropy. **b** Rescaled lateral pressure within mean-field theory (*solid line*) compared with the asymptotic infinite-stretching result (*dotted line*)

ever, all these theoretical predictions do not compare well with experimental results for the surface pressure of a compressed brush [12] which has to do with the fact that experimentally, chains are rather short so that one essen-

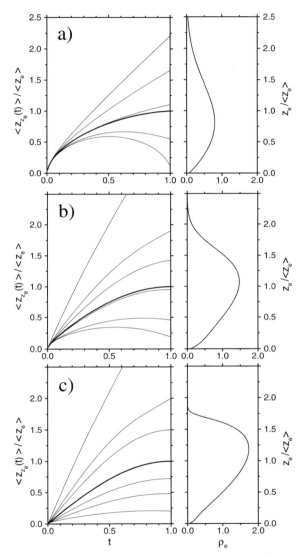

Fig. 5 *Left*: Mean-field results for the rescaled averaged polymer paths which end at a certain distance z_e from the wall for $\beta = 1, 10, 100$ (from *top* to *bottom*), corresponding to stretching values of $\gamma = 1.1, 1.9, 5.6$ (as defined in Eq. 18). The *thick solid line* shows the unconstrained mean path obtained by averaging over all end-point positions. Note that the end-point stretching is small but finite for all finite stretching parameters β. *Right*: End-point distributions

tially is dealing with a crossover situation between the mushroom and brush regimes. An alternative theoretical method to study tethered chains is the so-called single-chain mean-field method [53], where the statistical mechanics of a single chain is treated exactly, and the interactions with the other chains are taken into account on a mean-field level. This method is especially useful for short chains, where fluctuation effects are important, and for dense systems, where excluded volume interactions play a role. The calculated profiles and brush heights agree very well with experiments and computer simulations. Moreover, these calculations explain the pressure isotherms measured experimentally [12] and in molecular-dynamics simulations [57].

A further interesting question concerns the behavior of individual polymer paths. As was already discussed for the infinite-stretching theories ($\beta \to \infty$), polymers paths do end at any distance from the surface. In the left part of Fig. 5 we show mean-field results for the rescaled averaged polymer paths which end at a certain distance z_e from the wall for $\beta = 1, 10, 100$ (from top to bottom). Analyzing the polymer paths which end at a common distance from the surface, two rather unexpected features are obtained: (i) free polymer ends, in general, are stretched; and, (ii) the end-points lying close to the substrate are pointing towards the surface (such that the polymer path first turns away from the grafting surface before moving back towards it). In contrast, end-points lying *beyond* a certain distance from the substrate, point away from the surface (such that the paths move monotonously towards the surface). As we will explain shortly below, these two features have been confirmed in molecular-dynamics simulations [58]. They are not an artifact of the continuous self-consistent theory used in [52] nor are they due to the neglect of fluctuations. These are interesting results, especially since it has been long assumed that free polymer ends are unstretched, based on the assumption that no forces act on free polymer ends. The thick solid line shows the unconstrained mean path obtained by averaging over all end-point positions. Note that the end-point stretching is small but finite for all finite stretching parameters β. The right panel exhibits the end-point distributions, which, obviously, are finite over the whole range of vertical heights.

3.3
Molecular Dynamics Simulations

We now present results from molecular dynamics simulations in which all the chain monomers are coupled to a heat bath. The chains interact via the repulsive portion of a shifted Lennard-Jones potential with a Lennard-Jones diameter σ, which corresponds to a good solvent situation. For the bond potential between adjacent polymer segments we take a FENE (nonlinear bond) potential which gives an average nearest-neighbor monomer-monomer separation of typically $a \approx 0.97\sigma$. In the simulation box with a volume $L \times L \times L_z$ there are 50 (if not stated otherwise) chains each of which consists of $N + 1$

monomers with varying monomer number $N = 20, 30, 50$, as indicated. The box length L in the x- and y-direction was chosen to give anchoring densities ρ_a from $0.02\sigma^{-2}$ to $0.17\sigma^{-2}$. The first monomer of each chain is firmly and randomly attached to the grafting surface at $z = 0$. The mean square end-to-end distance of identical free (i.e. unanchored) chains was for the case $N = 50$ found to be $R_F \equiv \langle (r_N - r_0)^2 \rangle_{free}^{1/2} = 12.55\sigma$. For more details consult [58].

Figure 6 shows the behavior of the reduced monomer density $\rho(z)R_F/N\rho_a$ at increasing anchoring density. The stretching of the chains with increasing surface coverage, which is due to the repulsion between monomers, is evident. This plot has to be compared with Fig. 3b, where the same type of rescaling has been used. However, note that at this point, direct and quantitative comparison is not possible, since it is a priori not clear which value of the interaction parameter β in the self-consistent calculation corresponds to which set of simulation parameters σ, N, ρ_a.

The theoretically predicted scaling law of the brush height $h_0 \sim N\rho_a^{1/3}$ has been confirmed in several simulations [49]. Provided ρ_a is above the critical overlap density $\rho_a^* \sim N^{-6/5}$, the brush height, as measured by the first moment of the monomer density distribution $\langle z \rangle = \int z\rho(z)dz / \int \rho(z)dz$, should approach the predicted scaling form. Figure 7 shows this behavior in an appropriate scaling plot of the average brush height $\langle z \rangle$ and the z-component of the radius of gyration $R_G \equiv \langle \Sigma_{i=0}^{N}(r_i - r_{cm})^2/(N + 1) \rangle^{1/2}$ where r_{cm} is the position of the center of mass. For the scaling plot the brush height and radius of gyration are divided by the scaling prediction $h_0 \sim N\rho_a^{1/3}a^{5/3}$ and plotted as a function

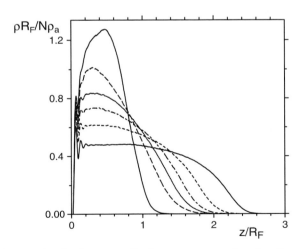

Fig. 6 Simulation results for the normalized monomer number density, $\rho(z)R_F/N\rho_a$, as a function of the scaled distance from the grafting surface z/R_F for anchored chains of length $N = 50$ and grafting densities $\rho_a\sigma^2 = 0.02, 0.04, 0.06, 0.09$, and 0.17 (from *top* to *bottom*). Note that R_F is determined within the simulation for a single, free polymer chain

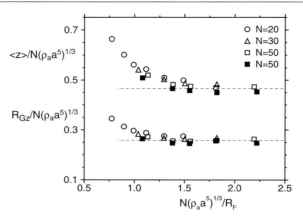

Fig. 7 Simulation results for the rescaled average monomer height $h/(N\rho_a^{1/3}a^{5/3})$ *(top)* and the vertical component of the radius of gyration $R_{Gz}/(N\rho_a^{1/3}a^{5/3})$ *(bottom)* versus the scaling variable $N(\rho_a a^5)^{1/3}/R_F$ for different chain lengths as indicated in the figure. The *open/filled symbols* correspond to extensible and nonextensible chains, respectively, which is controlled within the simulation by the strength of the FENE bond potentials. As one can see, for large values of the parameter combination $N(\rho_a a^5)^{1/3}/R_F$ the data saturate at a plateau and are thus in agreement with the scaling prediction

of the analogue of the interaction parameter, namely $h_0/R_F \sim N\rho_a^{1/3}a^{5/3}/R_F$, where a and R_F are determined from the simulation in the absence of a grafting wall. In this way we obtain an universal crossover point $N\rho_a^{1/3}a^{5/3}/R_F \approx 1.4$. Thus, for the chain length $N = 50$, the case we discussed most, one can be sure to reach the asymptotic scaling regime for the larger grafting densities.

A general problem when comparing experimental, simulation and analytical results among each other is that the different parameters have to be matched in a meaningful way. One such way is based on the relative stretching of chains in a brush. In Fig. 8 we plot in (a) simulation results for the stretching ratio

$$\gamma = \langle z_e(\rho_a)\rangle / \langle z_e(\rho_a = 0)\rangle , \qquad (18)$$

which is the averaged end-point height for a finite grafting density divided by the end-point height for vanishing grafting density. The stretching ratio is plotted as a function of the scaled anchoring density $(N\rho_a^{1/3}a^{5/3}/R_F)^2$ which is the analogue of the interaction parameter β. It is seen that the different chain lengths and grafting densities scale quite nicely. Clearly, for short chains or low grafting densities, that is in the mushroom regime, the stretching ratio γ approaches unity. The highest stretching ratio reached in the simulations is $\gamma \approx 2.7$. In Fig. 8b we show mean-field results for the stretching ratio γ as a function of the interaction parameter β. The general shape of the curve is similar. Matching the stretching ratios of the simulation and

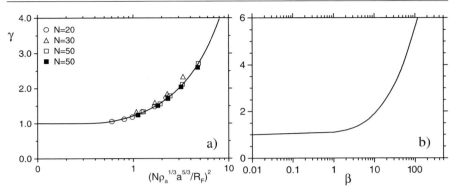

Fig. 8 Effective stretching factor $\gamma = \langle z_e(\rho_a)/z_e(0)\rangle$ as obtained **a** within the simulation as a function of the scaled anchoring density $(N\rho_a^{1/3}a^{5/3}/R_F)^2$ and **b** from the self-consistent field theory as a function of the interaction parameter β. The comparison between the simulation results and the analytical results allows us to determine the effective β parameter of a particular simulation set

the mean-field calculation allows us to determine the effective β value corresponding to a particular simulation run. As an example, for the most highly stretched simulation with $\gamma \approx 2.7$, we estimate an effective interaction parameter of $\beta \approx 25$. A similar matching procedure is possible with experimental data.

In agreement with the mean-field profiles shown in Fig. 3, the monomer density obtained from the simulations in Fig. 6 decays smoothly to zero at distances far from the anchoring plane. However, comparing the shape of the profiles in more detail, one notes a region over which the simulation profiles become rather flat for anchoring densities $\rho_a > 0.1\sigma^{-2}$, in contrast to the mean-field results. According to Fig. 8, this corresponds to stretching factors $\gamma > 2$ or $\beta > 10$. This discrepancy can be traced back to the neglect of higher-order virial terms in the analytical theory, which become important at elevated monomer densities.

We also looked at individual polymer paths within the simulation. The scaled average polymer paths with a given fixed end-point distance z_e, $\langle z_{z_e}(n)\rangle/\langle z_e\rangle$, are shown in Fig. 9 as a function of the rescaled contour variable n/N for three typical for three typical anchoring densities corresponding to stretching factors $\gamma \approx 1.3$, 1.8, and 2.7, from top to bottom. To get an idea of their relative weight we plot the polymer paths together with the normalized density of free ends $\rho_e(z)$ (shown in the right panel). The trajectories are very similar to those predicted by the mean-field theory (see Fig. 5). Paths which end far from the anchoring surface are stretched through their entire length, including the free end point. The paths which end at the outer rim of the distribution are almost uniformly stretched and appear almost as a straight line. On the other hand, paths which end close to the anchoring surface are nonmonotonic, first moving away from the wall, reaching a max-

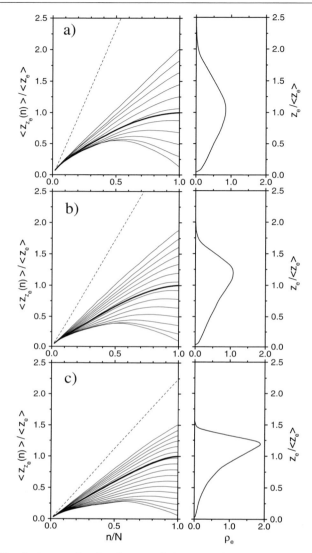

Fig. 9 *Left*: Simulation results for the rescaled averaged polymer paths which end at a certain distance z_e from the wall for $\rho_a \sigma^2 = 0.02, 0.06, 0.17$ (from *top* to *bottom*), corresponding to stretching values of $\gamma = 1.3, 1.8, 2.7$. The *thick solid line* shows the unconstrained mean path obtained by averaging over all end-point positions. *Right*: End-point distributions. All data are obtained for simulations with 50 chains consisting of $N = 50$ monomers each

imal distance, and then turning back towards the anchoring surface. Except at the maximum, all paths are stretched everywhere, including the end point. The straight broken lines are added for comparison and denote the maximally stretched possible path, which consists of a fully oriented and straight polymer. As can be seen from Fig. 5 as well as from Fig. 9, the end-point stretching, proportional to (dz/ds) or (dz/dn), respectively, is positive for some particular paths and negative for others, so that the average, plotted as a thick line in the figures, is typically quite small. This reconciles the present results with the infinite-stretching results by Milner, Witten and Cates [50] and Zhulina and co-workers [51], where the chains were assumed to be unstretched at their ends regardless of their end-point position: this assumption turns out to be acceptable on average if the stretching of the chains is large, i.e., for large β.

3.4
Additional Effects

As we described earlier, the main interest in end-adsorbed or grafted polymer layers stems from their ability to stabilize surfaces against van der Waals attraction. The force between colloids with grafted polymers is repulsive if the polymers do not adsorb on the grafting substrates [59], that is, in the absence of polymer bridges and loops. A stringent test of brush theories was possible with accurate experimental measurements of the repulsive interaction between two opposing grafted polymer layers using a surface force apparatus [8]. The resultant force could be fitted very nicely by the infinite-stretching theory of Milner et al. [60]. It was also shown that polydispersity effects, as appear in experiments, have to be taken into account theoretically in order to obtain a good fit of the data [61].

So far we assumed that the polymer grafted layer is in contact with a good solvent. In this case, the grafted polymers try to minimize their mutual contacts by stretching out into the solvent. If the solvent is bad, the monomers try to avoid the solvent by forming a collapsed brush, the height of which is considerably reduced with respect to the good-solvent case. It turns out that the collapse transition, which leads to phase separation in the bulk, is smeared out for the grafted layer and does not correspond to a true phase transition [62]. The height of the collapsed layer scales linearly in $\rho_a N$, which reflects the constant density within the brush, in agreement with experiments [63]. Some interesting effects have been described theoretically [64] and experimentally [63] for brushes in mixtures of good and bad solvent, which can be rationalized in terms of a partial solvent demixing.

For a theta solvent ($v_2 = 0$) the relevant interaction is described by the third virial coefficient; using a simple Alexander approach similar to the one leading to Eq. 13, the brush height is predicted to vary with the grafting density as $h \sim \rho_a^{1/2}$, in agreement with computer simulations [65].

Up to now we discussed planar grafting layers. It is of much interest to consider the case where polymers are grafted to *curved* surfaces. The first study taking into account curvature effects of stretched and tethered polymers was done in the context of star polymers [24]. It was found that chain tethering in the spherical geometry leads to a universal density profile, showing a densely packed core, an intermediate region where correlation effects are negligible and the density decays as $\rho(r) \sim 1/r$, and an outside region where correlations are important and the density decays as $\rho \sim r^{-4/3}$, with the radial distance denoted by r. These considerations were extended using the infinite-stretching theory of Milner et al. [66], self-consistent mean-field theories [67], and molecular dynamics simulations [68]. Of particular interest is the behavior of the bending rigidity of a polymer brush, which can be calculated from the free energy of a cylindrical and a spherical brush and forms a conceptually simple model for the bending rigidity of a lipid bilayer [69].

The behavior of a polymer brush in contact with a polymeric solvent, consisting of chemically identical but somewhat shorter chains than the brush, had been first considered by de Gennes [22]. A complete scaling description has been given only recently [70]. One distinguishes different regimes where the polymer solvent is expelled to various degrees from the brush. A somewhat related question concerns the behavior of two opposing brushes brought closely together, and separated by a solvent consisting of a polymer solution [71, 72]. Here one distinguishes a regime where the polymer solution leads to a strong attraction between the surfaces via the ordinary depletion interaction, but also a high polymer concentration regime where the attraction is not strong enough to induce colloidal flocculation. This phenomenon is called colloidal restabilization [71].

Considering a mixed brush made of mutually incompatible grafted chains, a novel transition to a brush characterized by a lateral composition modulation was found [73]. Even more complicated spatial structures are obtained with grafted diblock copolymers [74]. Finally, we would like to mention in passing that these static brush phenomena have interesting consequences for the dynamic properties of polymer brushes [75].

4
Charged Grafted Polymers

Brushes can also be formed by charged polymers which are densely endgrafted to a surface; the resulting *charged brush* shares many of the features discussed above for neutral brushes, but qualitatively new properties emerge due to the presence of charged monomers and counterions in the brush. Charged brushes have been the focus of numerous theoretical [76–84] and experimental [85–89] studies. They serve as an efficient mean for preventing colloids in polar media (such as aqueous solutions) from flocculating and precipitating out of solution. This stabilization arises from steric (entropic) as well as electrostatic repulsion. A strongly charged brush is able to trap its own counterions and generates a layer of locally enhanced salt concentration [78]. It is thus less sensitive to the salinity of the surrounding aqueous medium than a stabilization mechanism based on pure electrostatics (i.e. without polymers). Compared to the experimental knowledge about uncharged polymer brushes, less is known about the scaling behavior of PE brushes. The thickness of the brush layer has been inferred from neutron-scattering experiments on end-grafted polymers [85] and charged diblock copolymers at the air–water interface [87].

Theoretical work on PE brushes was initiated by the works of Miklavic and Marcelja [76] and Misra et al. [77]. In 1991, Pincus [78] and Borisov, Birshtein and Zhulina [79] presented scaling theories for charged brushes in the so-called osmotic regime, where the brush height results from the balance between the chain elasticity (which tends to decrease the brush height) and the repulsive osmotic counterion pressure (which tends to increase the brush height). In later studies, these works have been generalized to poor solvents [80] and to the regime where excluded volume effects become important, the so-called quasi-neutral or Alexander regime [83].

4.1
Scaling Approach

An analytical theory for polyelectrolyte brushes relies on a number of simplifying assumptions. The full theoretical problem is intractable because the degrees of freedom of the polymer chains and the counterions are coupled by steric and long-ranged Coulomb interactions. It is important to note that the charged polymer by itself is not fully understood, therefore quite drastic simplifications are needed to tackle the more complicated system of polyelectrolytes end-grafted to a surface. Firstly, we will concentrate on polymeric systems with counterions and only briefly mention the effects of added salt towards the end of this section, which has been discussed extensively in the original literature. Secondly, we will write the total free energy per unit area and in units of $k_B T$,

$$F = F_{pol} + F_{ion} + F_{int} \tag{19}$$

as a sum of separate contributions from the polyelectrolytes, F_{pol}, contributions from the counterions, F_{ion}, and an electrostatic interaction term F_{int} which couples polymers and counterions. The schematic geometry of the brush system is visualized in Fig. 10: we assume that the charged brush is characterized by two length scales. The polymer chains are assumed to extend to a distance h from the grafting surface, the counterions in general form a layer with a thickness of d. Two different scenarios emerge: The counterions can either extend outside the brush, $d \gg h$, as shown in Fig. 10a, or be confined inside the brush, $d \approx h$ as shown in Fig. 10b. As we demonstrate now, case (b) is indicative of strongly charged brushes, and thus applicable to

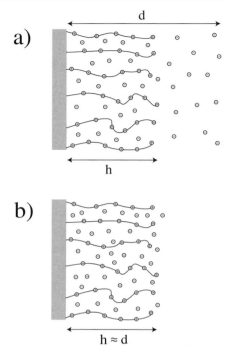

Fig. 10 Schematic PE brush structure. In **a** we show the weakly charged limit where the counterion cloud has a thickness d larger than the thickness of the brush layer, h. In **b** we show the opposite case of the strongly charged limit, where all counterions are contained inside the brush and a single length scale $d \approx h$ exists

most experiments done on charged brushes so far, while case (a) is typical for weakly charged brushes.

We recall that the grafting density of PE's is denoted by ρ_a, q is the counterion valency, N the polymerization index of grafted chains, and f the charge fraction. The counterion free energy contains entropic contributions (due to the confinement of the counterions inside a layer of thickness d) and also energetic contributions which come from interactions between counterions. In previous theories, a low-density expansion for the interaction part was used. We remind the reader that the second virial interaction is important for neutral brushes and is the driving repulsive force balancing the chain elasticity. For charged brushes, on the other hand, the leading term of the electrostatic correlation energy (which shows fractional scaling with respect to the charge density) is attractive and has been shown to lead to a collapse of the polyelectrolyte brush for large Bjerrum lengths [90, 91]. We will not pursue this collapse transition further in this review paper but solely concentrate on entropic and steric counterion effects, which is an acceptable approximation for mono-valent counterions. In the present analysis we use a free-volume approximation very much in the spirit of the van der Waals equation for the liquid–gas transition. We include the effective hard-core volume of a single polyelectrolyte chain, which we call V, and which reduces the free volume that is available for the counterions. This free volume theory therefore takes the hard-core interactions between the polymer monomers and the counterions into account in a nonlinear fashion and is valid even at large densities in the limit of close-packing. Compared to that, the excluded-volume interaction between counterions is small since the monomers are more bulky than the counterions and therefore it is neglected. The nonlinear entropic free energy contribution of the counterions reads per unit area

$$F_{\text{ion}} \simeq \frac{Nf\rho_{\text{a}}}{q} \left[\ln \left(\frac{Nf\rho_{\text{a}}/q}{d - \rho_{\text{a}}V} \right) - 1 \right].$$ (20)

In the limit of vanishing polymer excluded volume, $V \to 0$, one recovers the standard ideal entropy expression. As the volume available for the counterions in the brush, which per polymer is just d/ρ_{a}, approaches the self volume of the polymers, V, the free energy expression Eq. 20 diverges, that means, the entropic prize for that scenario becomes infinitely large. The excluded volume of the polymers is roughly independent of the polymer brush height, and can be written in terms of the effective monomer hardcore diameter σ_{eff} and the polymer contour length aN as $V \approx aN\sigma_{\text{eff}}^2$. This leads to the final expression

$$F_{\text{ion}} \simeq \frac{Nf\rho_{\text{a}}}{q} \left[\ln \left(\frac{Nf\rho_{\text{a}}/q}{d - Na\rho_{\text{a}}\sigma_{\text{eff}}^2} \right) - 1 \right].$$ (21)

The polymer free energy F_{pol} per unit area can in principle be derived from the free energy per chain used for neutral brushes, Eq. 12, by multiplying with the grafting density. It contains two contributions, one due to the chain elasticity and the other due to monomer-monomer interactions. Note that a logarithmic entropy term as for the counterions is missing, since the translational degrees of freedom of the monomers are lost due to chain connectivity. We recall that the entropy of chain stretching was accounted for by linear elasticity theory, with a spring constant proportional to $1/R_0^2$ in units of the thermal energy. This linear expression for the chain stretching is reliable for the neutral case, where indeed chain stretching is typically quite mild. For charged chains, the stretching is much stronger and we have to consider a more detailed model for the chain elasticity. For a freely jointed chain, which is a good model for synthetic polymer chains, the entropy loss due to stretching can be calculated exactly [92]. We only need here the asymptotic expressions for weak and for strong stretching, which read (per unit area)

$$F_{\text{pol}}^{\text{el}} \simeq \begin{cases} 3\rho_{\text{a}}h^2/(2Na^2) & \text{for} \quad h \ll Na \\ -\rho_{\text{a}}N\ln(1 - h/Na) & \text{for} \quad h \to Na \end{cases}$$ (22)

and are proportional to the grafting density ρ_a. The contour length of the fully stretched chain is Na. The weak-stretching term is the standard term used in previous scaling models. For the highly stretched situations encountered in fully charged brushes, the strong-stretching term is typically more appropriate and leads to a few changes in the results as will be explained further below. The energetic contribution to the polymer free energy can be expressed as the second virial contribution, arising from steric repulsion between the monomers (contributions due to counterions are neglected). Throughout this section, the polymers are assumed to be in a good solvent (positive second virial coefficient $v_2 > 0$). The contribution thus reads

$$F_{\text{pol}}^{\text{en}} \simeq \frac{1}{2} h v_2 \left(\frac{N \rho_a}{h} \right)^2 . \tag{23}$$

An additional electrostatic component to the polymer interaction term is typically unimportant since the counterions strongly screen any Coulomb interactions [92]. Finally, an electrostatic interaction between polymers and counterions F_{int} occurs if the PE brush is not locally electro-neutral throughout the system, an example is depicted in Fig. 10a. This energy is given by

$$F_{\text{int}} = \frac{2\pi \ell_B (N f \rho_a)^2}{3} \frac{(d - h)^2}{d} . \tag{24}$$

This situation arises in the limit of low charge, when the counterion density profile extends beyond the brush layer, i.e. $d > h$.

The different free energy contributions lead, upon minimization with respect to the two length scales h and d, to different behaviors. Let us first consider the weak charging limit, i.e. the situation where the counterions leave the brush, $d > h$. In this case, minimization of $F_{\text{ion}} + F_{\text{int}}$ with respect to the counterion height d in the limiting case of vanishing brush height ($h = 0$) and monomer volume ($\sigma_{\text{eff}} = 0$) leads to

$$d = \frac{3}{2\pi q \ell_B N f \rho_a} = 3\lambda_{\text{GC}} , \tag{25}$$

which has the same scaling as the so-called Gouy–Chapman length λ_{GC}. This length scale is a measure for the average height of the diffuse layer of q-valent counterions adsorbed at a surface with effective surface charge density $N f \rho_a$ and has been determined within simulations and field-theory [94] but can also be obtained with more coarse-grained scaling arguments, as demonstrated here. Balancing now the polymer stretching free energy $F_{\text{pol}}^{\text{el}}$ and the electrostatic energy F_{int} one obtains the so-called Pincus brush height

$$h \simeq N^3 \rho_a a^2 \ell_B f^2 , \tag{26}$$

which results from the electrostatic attraction between the counterions and the charged monomers. One notes the peculiar dependence on the polymerization index N. In the limit of $d \approx h$, the PE brush can be considered as neutral and the electrostatic energy vanishes. There are two ways of balancing the remaining free energy contributions. The first is obtained by comparing the osmotic free energy of counterion confinement, F_{ion}, in the limit when $d = h$ and for vanishing polymer volume, with the polymer stretching term, F_{pol}^{el}, in the weak stretching limit, leading to the height

$$h \sim \frac{Naf^{1/2}}{(3q)^{1/2}}, \tag{27}$$

constituting the linear osmotic brush regime. The main assumption here is that all counterions stay strictly localized inside the brush, which will be tested later by comparison with computer simulations.

Finally, comparing the second-virial term for the counterion interactions, F_{pol}^{en}, with the polymer stretching energy in the weak-stretching limit, F_{pol}^{el}, one obtains the same scaling behavior as the neutral brush [21, 22], compare Eq. 13. Comparing the brush heights in all three regimes we arrive at the phase diagram shown in Fig. 11. The three scaling regimes coexist at the characteristic charge fraction

$$f^{co} \sim \left(\frac{qv_2}{N^2 a^2 \ell_B} \right)^{1/3}, \tag{28}$$

and the characteristic grafting density

$$\rho_a^{co} \sim \frac{1}{N\ell_B^{1/2} v_2^{1/2}}. \tag{29}$$

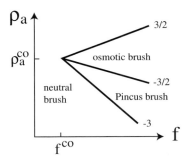

Fig. 11 Scaling diagram for PE brushes on a log-log plot as a function of the grafting density ρ_a and the fraction of charged monomers f. Featured are the Pincus-brush regime, where the counter-ion layer thickness is much larger than the brush thickness, the osmotic-brush regime, where all counterions are inside the brush and the brush height is determined by an equilibrium between the counter-ion osmotic pressure and the PE stretching energy, and the neutral-brush regime, where charge effects are not important and the brush height results from a balance of PE stretching energy and second-virial repulsion. The power law exponents of the various *lines* are denoted by numbers

4.2
Nonlinear Effects

The scaling relations for the brush height and the crossover boundaries between the various regimes constitute the simplest approach towards charged brushes. We have already pointed out a few limitations of the presented results, which have to do with nonlinear stretching and finite-volume effects.

Computer simulations provide an excellent mean to test those scaling relations. Extensive molecular dynamics simulations have been performed recently for polyelectrolyte brushes at various grafting densities and charge fractions, both at strong and intermediate electrostatic couplings [95, 96]. In these simulations, a freely-jointed bead-chain model is adopted for charged polymers end-grafted onto a rigid surface. The counterions are explicitly modeled as charged particles where both counterions and charged monomers are univalent and interact with the bare Coulomb potential (Eq. 8). The strength of the Coulomb interaction is controlled by the Bjerrum length ℓ_B. No additional electrolyte is added. The short-range repulsion between all particles is modeled by a shifted Lennard-Jones (LJ) potential, characterized by the hard-core diameter σ, as was done for the neutral brush simulations in Sect. 3.3. The simulation box is periodic in lateral directions and finite in the z-direction normal to the anchoring surface at $z = 0$. We apply the techniques introduced by Lekner and Sperb to account for the long-range nature of the Coulomb interactions in a laterally periodic system [95]. To study the system in equilibrium we perform stochastic molecular simulations at constant temperature.

Simulated density profiles of monomers and counterions of the system in the normal direction are shown in Fig. 12 for the fully charged brush at several grafting densities and for a Coulomb coupling characterized by $\ell_B/\sigma = 1$, which is close to the coupling of monovalent ionic groups in water. As seen, both monomers and counterions follow very similar nearly-step-like profiles with uniform densities inside the brush, which increases with grafting density. These data show that the counterions are mostly confined in the brush layer and that the electroneutrality condition is satisfied locally. The simple explanation is that the Gouy–Chapman length is indeed very small and of the order or smaller than the monomer diameter. One can observe that the polyelectrolyte chains are stretched up to about 70% of their contour length, which is roughly $L \approx N\sigma = 30\sigma$, and thus their elastic behavior is far beyond the linear regime. Therefore, within the chosen range of parameters, the simulated brush is in the strong-charging (i.e., all counterions are confined within the brush) and strong-stretching limits; as we will show below, it exhibits a nonlinear osmotic scaling behavior, so the linear scaling description of the last section, leading to Eq. 27, has to be slightly modified. The average height of end-points of the chains, $\langle z_e \rangle$, is one of the quantities which can be directly measured in the simulations and is shown in Fig. 13 together with

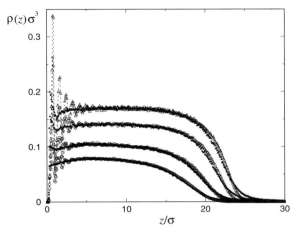

Fig. 12 Density profiles of monomers (*open symbols*) and counterions (*filled symbols*) as a function of the distance from the anchoring surface. Shown are profiles for fully charged brushes of 36 chains of $N = 30$ monomers at grafting densities (from *bottom* to *top*) $\rho_a \sigma^2 = 0.042$ (*circles*), 0.063 (*squares*), 0.094 (*diamonds*), and 0.12 (*triangles*). As can be clearly seen, the counterions stay inside the brush for all considered grafting densities and the local electroneutrality condition is satisfied very well

the predictions of the linear scaling theory, Eq. 27, denoted by (a), and the nonlinear scaling predictions, to be explained further below. It is observed that the simulated brush height (solid circles) varies slowly with the grafting density, contrary to the prediction of the linear scaling theory, Eq. 27, but in agreement with recent experimental results [97, 98].

We now present a scaling theory that incorporates nonlinear elastic and osmotic effects. In order to bring out the physics most clearly, the derivation will be done in two consecutive steps. First, in the strongly stretched osmotic brush regime, one chooses the strong-stretching version of the chain-stretching entropy in Eq. 22 and balances it with the counterion entropy, Eq. 21, assuming vanishing polymer self volume, $\sigma_{\text{eff}} = 0$, and assuming equal heights for the brush and counterion layers, $d = h$. The result is

$$h \sim \frac{Naf/q}{(1 + f/q)}, \tag{30}$$

which is the large-stretching analogue of Eq. 27. The maximal stretching predicted from this equation is obtained as f/q increases; for $f/q = 1$ one obtains a vertical chain extension corresponding to 50% of the contour length. For comparison, both expressions Eq. 27 and Eq. 30 are shown in Fig. 13 as dashed lines (a) and (b) respectively. Still, the overall stretching is considerably smaller than what is observed in experiments and simulations, and it transpires that something is missing in the above scaling description. It has to do with the entropic pressure which increases as the volume within the brush

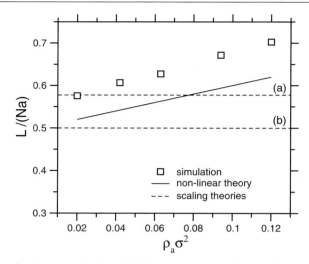

Fig. 13 The *circles* show simulation data for the brush height as a function of grafting density for polyelectrolyte chains of $N = 30$ monomers (contour length $L = 30a$) with a charge fraction of $f = 1$. The Bjerrum length is $\ell_B = \sigma$, which corresponds to an intermediate coupling strength quite relevant for fully charged chains in water, and all ions are univalent. The *dotted lines* (a) and (b) show the linear scaling predictions, Eqs. 27 and 30, with Gaussian and nonlinear elasticity respectively. The *solid line* shows the nonlinear scaling prediction according to Eq. 31, which includes nonlinear elasticity as well as the finite excluded volume of monomers and counterions

is progressively more filled up by monomers and counterions. Note that the brush height in Eq. 27 and Eq. 30 does not depend on the grafting density, in vivid contrast to the simulation results displayed in Fig. 13.

In the nonlinear osmotic brush regime we combine the high-stretching (nonlinear) version of the chain elasticity in Eq. 22 with the nonlinear entropic effects of the counterions due to the finite volume of the polymer chains, i.e. we choose a finite effective diameter $\sigma_{\rm eff}$ in Eq. 21. The final result for the equilibrium brush height is

$$h \sim \frac{Na\left(f/q + \sigma_{\rm eff}^2 \rho_a\right)}{(1 + f/q)}, \tag{31}$$

which in the limit of maximal grafting density, that is close packing, $\rho_a \to 1/\sigma_{\rm eff}^2$, reaches the maximal value stretching value, $h \to Na$, as one would expect; Compressing the brush laterally increases the vertical height and finally leads to a totally extended chain structure. In Fig. 13, we compare expression Eq. 31 with the simulation results for the brush height as a function of grafting density. Note that we have used $\sigma_{\rm eff}^2 = 2\sigma^2$. This choice corresponds to an approximate two-dimensional square-lattice packing of monomers and counterions on two interpenetrating sublattices. As can be seen, the scaling prediction, Eq. 31, qualitatively captures the slow brush height increase with grafting density, as observed in the simulations. The deviations between simulation data and the nonlinear osmotic brush prediction are unproblematic since the effective hard-core diameter of monomers in the simulation is difficult to determine precisely and might as well be treated as a fitting parameter. Deviations might also be explained by considering higher-order effects, such as lateral inhomogeneity of the counterion distribution around the brush chains and intermediate-stretching elasticity of the chains, that go beyond the present scaling analysis and have been considered in [92, 93].

4.3
Additional Effects

For large values of the charge fraction f and the grafting density ρ_a, and for large Coulomb coupling, it has been found numerically that the brush height does not follow any of the scaling laws discussed here [95]. This has been recently rationalized in terms of another scaling regime, the collapsed regime. In this regime one finds that correlation and fluctuation effects, which are neglected in the discussion in this section, lead to a net attraction between charged monomers and counterions [90, 91].

If salt is present in the solution, counterions as well as co-ions do penetrate into the brush, which leads to additional screening of the Coulomb repulsion inside the brush. The amount of this screening, and the stretching of the polyelectrolyte chains, are now also controlled by the bulk salt concentration. Since the additional salt screening weakens the swelling of the brush caused by the counterion osmotic pressure, salt leads to a brush contraction for sufficiently high salt concentration according to $h \sim c_{salt}^{-1/3}$ [78, 79]. The threshold salt concentration above which the brush contraction sets in is given by the salt concentration which equals the counterion concentration inside the brush. This means that the higher the grafting density (and consequently the higher the internal counterion concentration in the osmotic brush regime), the larger the salt concentration necessary to see any salt effects at all.

Another way of creating a charged brush is to dissolve a diblock copolymer consisting of a hydrophobic and a charged block in water. The diblocks associate to form a hydrophobic core, thereby minimizing the unfavorable interaction with water, while the charged blocks form a highly charged corona or brush [99]. The micelle morphology depends on different parameters. Most importantly, it can be shown that salt acts as a morphology switch, giving rise to the sequence spherical, cylindrical, to planar micellar morphology as the salt concentration is increased [99]. Theoretically, this can be explained by the entropy cost of counterion confinement in the charged corona [100]. The charged corona can be studied by neutron scattering [101] or atomic force microscopy [102] and gives information on the behavior of highly curved charged brushes.

5
Concluding Remarks

In this short review about neutral and charged polymers that are terminally grafted with one end to a surface we tried to explain and contrast the most important theoretical methods used for their understanding and description. The theories used for brushes are quite different from ordinary polymer modeling since the statistics of the grafted layer depends crucially on the fact that the chain is not attracted to the surface but is forced to be in contact to the surface since one of its ends is chemically or physically bonded to the surface. We review scaling concepts, mean-field theory and simulation techniques that give information on brushes at different levels of detail and accuracy. Scaling concepts yield the qualitative features of brushes, i.e. the brush height and its free energy as a function of the main parameters and without reliable prefactors. Mean-field or self-consistent theories allow us to construct density distributions on a coarse grained level, but correlation effects are completely missed [103]. Simulation techniques give in principle exact numerical results for the case of so-called primitive models where the solvent is neglected and monomers and ions are modeled as charged soft spheres. Current open questions concern the structure of water in such dense systems: is water still described by a continuous medium with the bulk dielectric constant? Secondly, all approaches neglect the varying polarizability of the monomers and ions, which again give rise to pronounced image-charge effects in the brush geometry. Lastly, most brush systems have not reached their true equilibrium structure, so that we are in essence dealing with a nonequilibrium system. How do we describe such nonequilibrium systems, and what governs the approach towards equilibrium including hydrodynamic and local friction effects? We believe that the triad of scaling, field-theoretic, and simulation techniques will also allow us to gain understanding of these more complicated issues in the near future.

References

1. Yamakawa H (1971) Modern Theory of Polymer Solutions. Harper & Row, New York
2. de Gennes PG (1979) Scaling Concepts in Polymer Physics. Cornell University, Ithaca
3. Grosberg AY, Khokhlov AR (1994) Statistical Physics of Macromolecules. AIP Press, New York
4. Rubinstein M (2003) Polymer Physics. Oxford University Press, Oxford
5. Netz RR, Andelman D (2003) Physics Reports 380:1
6. Halperin A, Tirell M, Lodge TP (1992) Adv Pol Sci 100:31
7. Szleifer I, Carignano MA (1996) Adv Chem Phys XCIV:165
8. Taunton HJ, Toprakcioglu C, Fetters LJ, Klein J (1988) Nature 332:712; (1990) Macromolecules 23:571
9. Auroy P, Auvray L, Leger L (1991) Phys Rev Lett 66:719; (1991) Macromolecules 24:2523; (1991) Macromolecules 24:5158

10. Marques CM, Joanny JF, Leibler L (1988) Macromolecules 21:1051; Marques CM, Joanny JF (1989) Macromolecules 22:1454
11. Field JB, Toprakcioglu C, Dai L, Hadziioannou G, Smith G, Hamilton W (1992) J Phys II France 2:2221
12. Kent MS, Lee LT, Factor BJ, Rondelez F, Smith GS (1995) J Chem Phys 103:2320
13. Bijsterbosch HD, de Haan VO, de Graaf AW, Mellema M, Leermakers FAM, Cohen Stuart MA, van Well AA (1995) Langmuir 11:4467
14. Teppner R, Motschmann H (1998) Macromolecules 31:7467
15. Oosawa F (1971) Polyelectrolytes. Dekker, New York
16. Förster S, Schmidt M (1995) Adv Polym Sci 120:50
17. Barrat JL, Joanny JF (1996) Adv Chem Phys 94:1
18. Holm H, Joanny JF, Kremer K, Netz RR, Reineker P, Seidel C, Vilgis TA, Winkler RG (2004) Adv Polym Science 166:67
19. Rühe J, Ballauff M, Biesalski M, Dziezok P, Gröhn F, Johannsmann D, Houbenov N, Hugenberg N, Konradi R, Minko S, Motornov M, Netz RR, Schmidt M, Seidel C, Stamm M, Stephan T, Usov D, Zhang H (2004) Adv Polym Science 165:79
20. Dolan AK, Edwards SF (1974) Proc R Soc Lond A 337:509; 343:427
21. Alexander S (1977) J Phys (France) 38:983
22. de Gennes PG (1980) Macromolecules 13:1069
23. Semenov AN (1985) Sov Phys JETP 61:733
24. Daoud M, Cotton JP (1982) J Phys France 43:531
25. Karim A, Satija SK, Douglas JF, Ankner JF, Fetters LJ (1994) Phys Rev Lett 73:3407
26. Baltes H, Schwendler M, Helm CA, Heger R, Goedel WA (1997) Macromolecules 30:6633
27. Netz RR (2003) J Phys Condens Mat 15:S239
28. Friedsam C, Gaub HE, Netz RR (2005) Europhys Lett 72:844
29. Burak Y, Netz RR (2004) J Chem Phys B 108:4840
30. Odijk T (1977) J Polym Sci Polym Phys Ed 15:477; (1978) Polymer 19:989
31. Skolnick J, Fixman M (1977) Macromolecules 10:944
32. Barrat JL, Joanny JF (1993) Europhys Lett 24:333
33. Netz RR, Orland H (1999) Eur Phys J B 8:81
34. Manghi M, Netz RR (2004) Eur Phys J E 14:67
35. Ullner M, Woodward CE (2002) Macromolecules 35:1437
36. Everaers R, Milchev A, Yamakov V (2002) Eur Phys J E 8:3
37. Netz RR, Joanny JF (1999) Macromolecules 32:9013; Netz RR, Joanny JF (1999) Macromolecules 32:9026
38. Manning GS (1969) J Chem Phys 51:924
39. Manning GS (1969) J Chem Phys 51:934
40. Manning GS, Ray J (1998) J Biomol Struct Dyn 16:461
41. Naji A, Netz RR (2005) Phys Rev Lett 95:185703
42. Fixman M (1982) J Chem Phys 76:6346
43. Le Bret M (1982) J Chem Phys 76:6243
44. Manning GS, Mohanty U (1997) Physica A 247:196
45. Manning GS (1988) J Chem Phys 89:3772
46. Wandrey C, Hunkeler D, Wendler U, Jaeger W (2000) Macromolecules 33:7136
47. Blaul J, Wittemann M, Ballauff M, Rehahn M (2000) J Phys Chem B 104:7077
48. Cosgrove T, Heath T, van Lent B, Leermakers FAM, Scheutjens J (1987) Macromolecules 20:1692
49. Murat M, Grest GS (1989) Macromolecules 22:4054; Chakrabarti A, Toral R (1990) Macromolecules 23:2016; Lai PY, Binder K (1991) J Chem Phys 95:9288

50. Milner ST, Witten TA, Cates ME (1988) Europhys Lett 5:413; (1988) Macromolecules 21:610
51. Skvortsov AM, Pavlushkov IV, Gorbunov AA, Zhulina YB, Borisov OV, Pryamitsyn VA (1988) Polymer Science 30:1706
52. Netz RR, Schick M (1997) Europhys Lett 38:37; (1998) Macromolecules 31:5105
53. Carignano MA, Szleifer I (1995) J Chem Phys 102:8662
54. Martin JI, Wang ZG (1995) J Chem Phys 99:2833
55. Baranowski R, Whitmore MD (1995) J Chem Phys 103:2343
56. Currie EPK, Leermakers FAM, Cohen Stuart MA, Fleer GJ (1999) Macromolecules 32:487
57. Grest GS (1994) Macromolecules 27:418
58. Seidel C, Netz RR (2000) Macromolecules 33:634
59. Witten TA, Pincus PA (1986) Macromolecules 19:2509; Zhulina EB, Borisov OV, Priamitsyn VA (1990) J Coll Surf Sci 137:495
60. Milner ST (1988) Europhys Lett 7:695
61. Milner ST, Witten TA, Cates ME (1989) Macromolecules 22:853
62. Halperin A (1988) J Phys France 49:547; Zhulina EB, Borisov OV, Pryamitsyn VA, Birshtein TM (1991) Macromolecules 24:140; Williams DRM (1993) J Phys II France 3:1313
63. Auroy P, Auvray L (1992) Macromolecules 25:4134
64. Marko JF (1993) Macromolecules 26:313
65. Lai PY, Binder K (1992) J Chem Phys 97:586; Grest GS, Murat M (1993) Macromolecules 26:3108
66. Ball RC, Marko JF, Milner ST, Witten TA (1991) Macromolecules 24:693; Li H, Witten TA (1994) Macromolecules 27:449; Manghi M, Aubouy M, Gay C, Ligoure C (2001) Eur Phys J E 5:519
67. Dan N, Tirrell M (1992) Macromolecules 25:2890
68. Murat M, Grest GS (1991) Macromolecules 24:704
69. Milner ST, Witten TA (1988) J Phys France 49:1951
70. Aubouy M, Fredrickson GH, Pincus P, Raphael E (1995) Macromolecules 28:2979
71. Gast AP, Leibler L (1986) Macromolecules 19:686
72. Borukhov I, Leibler L (2000) Phys Rev E 62:R41
73. Marko JF, Witten TA (1991) Phys Rev Lett 66:1541
74. Brown G, Chakrabarti A, Marko JF (1995) Macromolecules 28:7817; Zhulina EB, Singhm C, Balazs AC (1996) Macromolecules 29:8254
75. Halperin A, Alexander S (1988) Europhys Lett 6:329; Johner A, Joanny JF (1990) Macromolecules 23:5299; Ligoure C, Leibler L (1990) J Phys France 51:1313; Milner ST (1992) Macromolecules 25:5487; Johner A, Joanny JF (1993) J Chem Phys 98:1647
76. Miklavic SJ, Marcelja S (1988) J Phys Chem 92:6718
77. Misra S, Varanasi S, Varanasi PP (1989) Macromolecules 22:5173
78. Pincus P (1991) Macromolecules 24:2912
79. Borisov OV, Birstein TM, Zhulina EB (1991) J Phys II (France) 1:521
80. Ross RS, Pincus P (1992) Macromolecules 25:2177; Zhulina EB, Birstein TM, Borisov OV (1992) J Phys II (France) 2:63
81. Wittmer J, Joanny JF (1993) Macromolecules 26:2691
82. Israels R, Leermakers FAM, Fleer GJ, Zhulina EB (1994) Macromolecules 27:3249
83. Borisov OV, Zhulina EB, Birstein TM (1994) Macromolecules 27:4795
84. Zhulina EB, Borisov OV (1997) J Chem Phys 107:5952
85. Mir Y, Auvroy P, Auvray L (1995) Phys Rev Lett 75:2863

86. Guenoun P, Schlachli A, Sentenac D, Mays JM, Benattar JJ (1995) Phys Rev Lett 74:3628
87. Ahrens H, Förster S, Helm CA (1997) Macromolecules 30:8447; (1998) Phys Rev Lett 81:4172
88. Ballauff M, Guo X (2001) Phys Rev E 64:051406
89. Balastre M, Li F, Schorr P, Yang J, Mays JW, Tirrell MV (2002) Macromolecules 35:9480
90. Csajka FS, Netz RR, Seidel C, Joanny JF (2001) Eur Phys J E 4:505
91. Santangelo CD, Lau AWC (2004) Eur Phys J E 13:335
92. Naji A, Netz RR, Seidel C (2003) Eur Phys J E 12:223
93. Hugel T, Rief M, Seitz M, Gaub HE, Netz RR (2005) Phys Rev Lett 94:048301
94. Moreira AG, Netz RR (2001) Phys Rev Lett 87:078301; (2002) Eur Phys J E 8:33
95. Csajka F, Seidel C (2000) Macromolecules 33:2728
96. Seidel C (2003) Macromolecules 36:2536
97. Romet-Lemonne, Daillant J, Guenoun P, Yang J, Mays JW (2004) Phys Rev Lett 93:148301
98. Ahrens H, Förster S, Helm CA, Kumar NA, Naji A, Netz RR, Seidel C (2004) J Chem Phys B 108:16870–16876
99. Shen H, Zhang L, Eisenberg A (1999) J Am Chem Soc 121:2728
100. Netz RR (1999) Europhys Lett 47:391
101. Guenoun P, Muller F, Delsanti M, Auvray L, Chen YJ, Mays JW, Tirrell M (1998) Phys Rev Lett 81:3872; Guenoun P, Delsanti M, Gazeau D, Mays JW, Cook DC, Tirrell M, Auvray L (1998) Eur Phys J B 1:77
102. Förster S, Hermsdorf N, Leube W, Schnablegger H, Regenbrecht M, Akarai S, Lindner P, Böttcher C (1999) J Phys Chem B 103:6657
103. Boroudjerdi H, Kim YW, Naji A, Netz RR, Schlagberger X, Serr A (2005) Phys Rep 416:129

Author Index Volumes 101–198

Subject Index

Printing: Krips bv, Meppel
Binding: Stürtz, Würzburg